A BRIEF TREATISE ON GEOLOGY;

OR

FACTS, SUGGESTIONS,

AND

INDUCTIONS IN THAT SCIENCE.

BY

BIBLICUS DELVINUS.

SECOND EDITION.

INTRODUCTION.

Whoever enters deeply into the study of Natural Philosophy, with an unbiassed and devotional turn of mind, will perceive that there exists a remarkable connexion between the scriptural records of such philosophy, and the discoveries of modern science, especially in those relating to Geology. As in the prophetic declarations of the Bible, the truth of God's word has been demonstrated by clearer discoveries of the actual fulfilment of those prophecies, so science, if rightly applied will elucidate and become as it were the handmaid and illustrator of divine revelation, leading us to repose an implicit trust in the word of that Being, who in the beginning created all things, and pronounced them " very good."

From the mode at present adopted, the task appears replete with difficulties to reconcile the discoveries of Modern Geology with the Mosaic detail of the creation; but these difficulties are increased by the Geologist himself, for when he arrives at the knowledge of what he considers *a fact* in Geology, in most cases he is unwilling to bring such a fact, like gold to the touchstone of all truth, in order to see, what light divine revelation will throw upon his discovery.

It was the opinion of Lord Verulam, that the Book of Job is full of the mysteries of natural philosophy; and other philosophers have held, that there is a complete system of Geology obscurely shadowed forth in this book, and in the first chapter of Genesis, though man may not always succeed in gathering it from the incidental and elementary passages in which it is embodied.

In the first Chap. of Genesis it is revealed *that God created the heavens and the earth;* and a formal account of the work is there particularly related.

But according to the opinion of some Geologists, God is an insignificant word, and the account of the creation there given is false; so that every one is still left at liberty to frame what notions he pleases, and to lay down his own laws for the laws of nature. Are not such men as much in the dark about these important points, as if nothing had been revealed? That Moses, the legislator of the Hebrews, and a prophet of God, should introduce his laws with a false account of the creation and formation of things; and that the God of truth should allow, much less inspire, his servant to hand down to posterity a formal untruth, is repugnant to common sense, and sounds harsh to a Christian's ear: yet what less does every one say in his heart, who denies the Mosaic philosophy to be true?

If Moses accommodated his account of natural things to the understanding of the vulgar, and if the Jews were so carnal a people as supposed, might he not do the same with regard to spiritual things? So that this supposition, of writing for the vulgar, by clearing the inspired record of one objection, láys it open to another, viz. That if he conformed or adapted his account to their conception of nature, he might do the like in respect to God. And if so, it is to no purpose to look for either sound philosophy or divinity in his writings. It may be objected, however, that God, by his revelation, did not intend to instruct men either in Geology or Philosophy generally, but in faith and obedience. Had this been the case, revelation would have been confined to those points solely, or at least chiefly, and not have so largely and frequently intermixed an account of physical phenomena with them; so that if all the places of scripture which treat of, or relate to philosophical subjects, were collected together, they would make one half or nearly two thirds of the Bible. Besides, when God was rectifying the errors of their conceits with regard to his nature and worship, and giving them a complete ritual to preserve them in the true faith and worship, would he let them remain in all their errors and false opinions with regard to Nature and her operations? Nay, stamp them with his authority, by treating of them in the same idle manner men were at that time wont to conceive of them? This is to make the Holy Ghost lie unto man. And if the inspired writers of the books of the Old Testament understood the real nature of things, and their true causes, as it is expressly asserted of Solomon, then by this supposition, did the Holy Ghost over-rule these penmen to suppress their real knowledge of nature, and to transmit down to posterity under their hands what they knew to be false? To such strange absurdities does this opinion drive its advocates. If the knowledge of philosophy or nature be attainable by human reason and application, there was no need of a revelation: yet a revelation has been given. Why must it be adapted only to the vulgar? A revelation being made, and that from God, implies it to be, in the nature of things, both necessary, full, perfect, and true. Otherwise why was it made that after ages should find it to be false? This cannot be supposed, without highly derogating from the wisdom and goodness of the supreme Being; and yet it is supposed even by some of our learned men. They say, That philosophy not being necessary to salvation, and being what man could by his natural abilities discover, the inspired writers did not meddle with it; and what we meet with in the Scriptures, being only incidentally spoken, they conformed to the received opinions of the persons they wrote to; and that as to the Deity, the New Testament has fully opened his nature, essence, and personality, which were either wholly unknown, or but faintly revealed to the Jews.

Whether philosophy is necessary to salvation we need not now stop to inquire: but when we know what philosophy is, it may, in its proper place, be shewn to be more necessary to theology than is generally thought; and that human reason with all its boasted discoveries, has not yet been able to find out the true system of the world. Indeed, when matters are fairly stated, and rightly understood, it will appear that the Jews were not so ignorant either of *God* or *Nature* as some imagine who do not know the Scriptures and the power of God.

That philosophy is incidental only, is an assertion not correct. The books of Moses begin with an account of the creation, and the steps the Divine Being took in establishing this mundane system.

There is also mention made of the manner in which the earth was destroyed by a flood, and of its being restored again to an inhabitable state. And those accounts are so circumstantial, that they seem designed rather to rectify than to conform to the mistaken notions of mankind; and shew, in the clearest manner, that the mention of philosophy in the Scriptures is not incidental only nor accomodated merely to vulgar minds.

But why, it may be asked, should it be thought by some a thing incredible, that the Scriptures should give a true description of nature as well as of its great Author?* Could not then the Creator of all things frame this world so as to give some idea or resemblance of himself? Would

---

* The miracle of Joshua commanding the sun to stand still, is no argument against this position, as is most satisfactorily proved by Jones in his "First Principles of Natural Philosophy and Theology," and Pikes' "Philosophia Sacra;" in which works the nature of that miracle and also various expressions referring to the operations of nature, are ably and fully discussed.

he order his servants to use names not significant, or improper, for his being and attributes? Or can we suppose a language formed by his direction, while Adam was perfect, vague and uncertain and to want words properly expressive of what he thought fit to reveal of himself and his works? Would he give a revelation defective in two such main points, at a time when he was separating the Jews from the rest of the nations, who had apostatized from the true God, *the Maker of the heavens and the earth,* to the worship of his *material agents?* The contest was then the same as in Elijah's time, Whether Jehovah was God, or Baal was God. Whether the sun, moon, stars, and all the host of heaven, were uncreated, eternal, and so God; or whether the Lord Jehovah, the God of the Jews, created and made the heavens, the earth, and all the host of heaven? Was a short dark account in words inexpressive of God's essence or powers, and a false unintelligible one of nature, a likely method to determine this contest? Why should God give any account of nature, if not a true one? Why direct his servants to write for vulgar understandings only? And if men were so ignorant as to swallow a false account, could not infinite wisdom foresee, that the present learned race would arise, and be framing schemes to detect the falsity of it? Was not the knowledge of God and nature as necessary to men *then* as to men *now?* And if God was pleased to record his revelation of these things, was he not bound by his veracity to do it truly and intelligibly? When our translation was made, the learned and unlearned both agreed in their notion of philosophy; so the translation was agreeable to the received opinion. But there has since been a progressive improvement in science, whereby men have obtained a more thorough knowledge of certain physical facts and phenomena; and had the translators of the Bible commenced their work in our times, their translation, (without supposing them to *bend* to the present notions), would have been more in accordance with the actual and true system of nature.

These and many such reflections, which naturally arise in the mind employed on this subject, lead us to conclude, from the necessity of the thing, and the wisdom and goodness of the Deity, that a full and perfect revelation was at first made, and afterwards renewed, as occasion served, when the wantonness of imagination had corrupted and defaced it, and at last committed to writing by Moses, for the use of the then present and all succeeding generations, to be the rule and standard whereby they should measure and regulate their opinions and belief of things natural and divine.

The heathen, we are told in Scripture, thought the sun, moon, and host of heaven, Gods; so worshipped them; and no doubt mocked the prophets who preached to them, That the Lord made the heavens and the earth. Pharaoh knew not the Lord, neither would obey his voice: why? because his magicians deluded him by experiments, into a belief that the powers which Moses attributed to the *Lord*, were lodged in the *material agents*. So now some of our Geologists, by digging into the bowels of the earth pretend to find out the works of God; and in endeavouring to *fix the age of the world*, have built a system of theology diametrically opposite, in the fundamental point, to what is taught in Scripture; but the day is coming it is to be hoped, when they will see that nothing but true philosophy is in keeping with divine revelation, for those who are prejudiced with false notions of the one, cannot be brought to entertain just and proper notions of the other.

It has been the Author's object in the present Treatise, to search into the workings of nature, and thence to display the *truth*, goodness, wisdom, and power of the Creator, which the wonders of Geology unfold. Whilst tracing out a scriptural system, his mind has been unbiassed by any one opinion however generally received, having found it easier to get nearer the truth by proposing an *accommodation* which is done in the fourth chapter, between the two antagonist theories, that of Werner and Doctor Hutton. Being anxious to lay the foundation of his remarks in well tried and undeniable experiments, he has not scrupled to avail himself of the practical researches of scientific men as contained in their works, but wishing as much as possible to avoid darkening counsel by words without knowledge, he has preferred condensing his Facts, Suggestions, &c., into a small compass, being satisfied, that if there is any truth contained in them, it is not the thickness or size of the book, which is to recommend them to the attentive consideration of the public.

The last chapter of this second edition contains some additional matter, with observations on the formation of valleys, and an explanation of the phenomena of those erratic blocks which are found scattered at a distance from their parent rocks in every country of the globe.

CONTENTS.

CHAP. I.

PAGE

Form of our earth at the creation—Design of the Creator in thus forming it—Salt an original formation; its origin 8

CHAP. II.

Salt and Organic Remains peculiar to Basins—A more general research necessary to establish any Geological Theory—How to discover if there have been successive formations—The Scriptures will reflect light on Geology—The design considered of the progression from simple to more complex forms—Molecules, the nutriment provided for the earliest marine inhabitants—All organized and unorganized bodies composed of Molecules—The place of Molecular origin—Balance of Molecular Affinities—Cause and period of the derangement of the balance of Affinities suggested by which the primitive Rocks were formed—Molecules the substratum of all substance—This confirmatory of divine truth . . 23

CHAP. III.

Much of the earth's surface formed by organic secretion—Remains not confusedly heaped together—No higher orders of organic beings originally deposited in the most ancient strata—In the upper strata organic remains are found confused and broken, cause explained—Mr. Brown's Experiment on Molecules—What Molecular motion and rest seem to indicate—Period and condition of our globe when many of the organic tribe perished—In what manner their remains are found—Total change of the primeval earth and atmosphere which accounts for the extinction of many of the organic beings—Salubrity of the primitive atmosphere most favourable to vegetable and organic life—Vegetable and organic life kept in due bounds by herbiverous and carniverous animals—Earthquakes the second causes of the Deluge—Manner in which many of the fossil fishes are found—Period and manner in which they were destroyed—Cause suggested why the human bones of the Antediluvian race are not to be found 41

CHAP. IV.

Period and manner considered in which the primitive mountains were formed—When some were elevated by Earthquakes—On the formation of the vast stratified Deposits and beds containing organic remains—Attempts made to account for the latter by supposing prolonged ages of Creation—This hypothesis not in accordance with Scripture—Period when some of the metallic veins were formed, and how—Majority of the strata, how formed—*Conflicting* views, how reconciled—The two theories of veins being filled from above and below, considered—Metallic ores which occur in beds, and regular strata when formed—Period and manner in which veins were filled from above and below—Metallic veins containing different ores at various depths, accounted for by the balance of affinities—Electricity an agent in the formation of some veins—The formation of some metallic veins has a close connexion with the period in which the Coal strata were deposited—Manner in which the Coal strata were formed—Faults in the coal beds—The beneficent design of the Creator in the arrangement of the minerals—Central heat—Earthquakes, and Volcanoes, confined to certain spaces—Difficulties that attend the theory of Central heat—Experiments to ascertain the tem. in Mines and Artesian Wells—Increase of the tem. ascribed to subterranean fires—Their action in the Mining districts and mountainous chains—An analogy between the tem. of the Thermal waters and the height of the mountains near which they occur—Tem. of the Thermal waters of England and of some on the Continent—General researches necessary to ascertain the tem. of Artesian wells—Boring for water at Boston in Lincolnshire, the result—To ascertain the tem. of Artesian wells, the borings should be through clay, the alluvial deposits, gypsum, and shale—Tem. of different mines—Remarkable variation—This at variance with the theory of a general central heat—The high tem. of the *water* in mines attributed to volcanic heat—Theories on the formation of vallies—On the phenomena of erratic blocks—How we should regard the convulsions which have taken place in the earth 55

PRELIMINARY OBSERVATIONS.

The following question must often present itself to the student in Geology: *Of what materials is the world we live in composed?* This question, with certain conditions, comprises the most important objects of geological research; namely—*What are the substances* of which the earth is composed? *What is the order in which they are arranged? And what are the changes they appear to have undergone?*—But how are satisfactory answers to these inquiries to be obtained?

From the surveys that have been made of the solid crust of the earth, so far as it has been penetrated, it is evident, that the rudimental materials of the globe existed, at its earliest period, in a liquid mass: that they were afterwards separated and arranged by a progressive series of operations, and an uniform system of laws; and that they have since been convulsed and dislocated by some dreadful commotion and inundation that have extended to every region, and again thrown a great part of the organic and inorganic creation into what appears to us to be a promiscuous jumble.

Hence, have originated the *Plutonic* and the *Neptunian* hypotheses: the former, ascribing the origin of the world, in its present state, to *igneous* fusion; the latter, to *aqueous* solution.

The Plutonic theory aspires, in many of its modifications, not only to account for the present appearances of the earth, but for that of the universe; and traces out a scheme by which every planet, or system of planets, may be continued indefinitely, and, perhaps, for ever, by a perpetual series of restoration and balance.

With this system the Neptunian forms a perfect contrast. It is limited to the earth, and the present appearances of the earth. It resolves the general origin of things into the operation of water; and while it admits the existence of subterranean fires to a certain extent, and that several of the phenomena which strike us most forcibly must be the result of such agency, it peremptorily denies, that this agency is the sole or universal cause of the existing state of things, or that it could possibly be rendered competent to such an effect.

Of these two theories, the Neptunian is more in accordance with the geology of the Scriptures. The Mosaic narrative, indeed, takes a bold and comprehensive sweep through the vast range of the solar system, if not through that of the universe; and in its history of the simultaneous origin of this system, touches chiefly on geology, as the part most interesting to ourselves; but so far as it enters upon this doctrine, it is in general coincidence with the Neptunian scheme, and with the great volume of Nature as now cursorily dipped into. The narrative opens with a statement of three distinct facts, each following the other in a regular series, in the origin of the visible world. First, an absolute creation, as opposed to a re-modification of the heaven and the earth, which constituted the *earliest* step in the productive process. Secondly, the condition of the earth when it was thus primarily brought into being, which was that of an amorphous or shapeless waste. And, thirdly, a commencing effect to reduce the unfashioned mass to a condition of order and harmony. " In the beginning," says the sacred historian, " God *created* the heaven and the earth. And the earth was *without form and void:* and darkness was on the face of the deep (or abyss.)—And the Spirit of God *moved* upon the *face of the waters.*"*

We are hence, therefore, necessarily led to infer, that the first change of the formless chaos, took place when the Spirit commenced his operative power by moving upon the surface of the waters. We are next informed that this chaotic mass acquired shape, not instantaneously, but by a series of six distinct days, each being of the same duration as what we *now* understand to be a day.

The Mosaic narrative proceeds to tell us, that during the first of these days was evolved, what agreeably to the laws of gravity, must have been evolved first of all the matter of light and heat, of all material substances the most subtle and

* If we look to what has been collected from the Heathens, we shall find that their traditions agree with the sacred account in the following particulars: first, that the earth was originally in a formless state; secondly, that its materials floated confusedly in a watery abyss; thirdly, that the matter of the heavens was *dark* and *stagnant;* fourthly, that the night was prior to the day, or that Erebus, or evening was before the day-light; fifthly, that the world was subject to a sort of incubation from the Spirit; and sixthly, that *water* was held most sacred, because it was the first of all things.

attenuate; those by which alone the sun operates, and has ever operated upon the earth and the other planets, and which may be the identical substances that constitute its essence. And it tells us, also, that the luminous matter thus evolved produced light without the assistance of the sun or moon, which were not set in the sky or firmament, and had no rule till the fourth day : that the light thus produced flowed by tides, and alternately intermitted, constituting a single day and a single night of each of the six periods or days of creation.

It tells us, that during the second day, uprose progressively the fine fluids, of the firmament, and filled the blue etherial void with a vital atmosphere. That, during the third day, the waters, more properly so called, or the grosser and compacter fluids of the general mass, were strained off and gathered together into the vast bed of the ocean, and the abyss in the hollow of the earth, and the dry land began to make its appearance, by discovering the peaks or highest points of the primitive mountains : in consequence of which, a progress instantly commenced from inorganic matter to vegetable organization, the surface of the earth, as well above as under the waters, being covered with plants and herbs bearing seeds after their respective kinds; thus laying a basis for those carbonaceous materials, the remains of vegetable matter, which are occasionally to be traced in some of the layers or formations of the class of primitive rocks, (the lowest of the whole,) without a single particle of animal relics intermixed with them.

It tells us, that during the fourth day, the sun and moon, now completed, were set in the firmament, the solar system was finished, its laws were established, and the celestial orrery was put into play; in consequence of which, the harmonious revolutions of signs and seasons, of days and of years, struck up for the first time their mighty symphony. That the fifth period was allotted exclusively to the formation of water fowl and the countless tribes of aquatic creatures. It tells us,—still continuing the same grand and exquisite climax—that towards the close of this period, the mass of waters having sufficiently retired " unto the place which God had founded for them," Psalm civ. 8. the sixth and concluding period was devoted to the formation of terrestrial animals; and, last of all, as the master-piece of the whole, to that of man himself.

Thus, in progressive order, uprose the stupendous system of the world : the bright host of morning stars shouted together on its birth-day; and the eternal Creator looked down with complacency on the finished fabric, and "saw that it was very good."

We have seen how Scripture teaches us the order in which the Divine Intelligence created our earth, and the truth of the foregoing detail is amply confirmed by the researches of modern geologists. But there have since been desolations, changes, and new formations effectuated by the universal deluge on the surface and upper strata of the earth, as well as fractures, upheaving, and subsidences caused by earthquakes, which it is exclusively the province of the geologist to notice.

In tracing out the designing intelligence of the Creator in various of his formations, it has been the author's object to endeavour also to answer the three questions proposed at the outset of these observations, namely, *What are the substances* of which the earth is composed? *What is the order in which they are arranged?* And, *what are the changes they appear to have undergone?*

CHAPTER I.

It has frequently been the remark of the author of this little treatise, that he has never yet met with any person, who either in writing or in conversation, seems to have made up his mind according to geological theory, as to the precise form or state, in which the earth must of necessity have existed before that period given in the Mosaic detail, the formation of light. It being requisite that every one should have clear and well defined notions on this subject, (especially if they wish to make Geology their study,) it is here proposed with the aid of Scripture to attempt determining, in as concise and clear a manner as possible, what was the *form* and *state* of the earth, before it arrived at its present condition.

In the Mosaic narrative we are told, that, " In the beginning, God created the heaven and the earth; and the earth was without form and void; and darkness was on the face of the deep; and the Spirit of God moved upon the face of the waters."

Many commentators agree, that between the second and third verses of the ancient Hebrew versions there exists an obvious break; by which it is evidently signified, that our earth *existed*, in some form or other, previous to the moment, when God said,—" Let there be light." Should any be unwilling to allow, that such a break does occur, it need not be insisted on; for we are plainly told in the second verse of the first chapter of Genesis, that the earth *did* once exist in a *different* state from that in which we now find it; only it was "without form and void." The first act of remodelment* from this shapeless

* The word remodelment is here used in preference, because matter such as it was in the foundation rocks, the deep, and darkness, existed *previously*—it was now that the first shaping process commenced, whereby the darkness was converted into light, or, to use the more emphatical Hebrew phrase, God divided between the light, and between the darkness, taking from one and giving to another, which, before the interruption, were mixed one with the other. Next was separated the firmament from the waters, and these gathered together to one place, whereby the dry land appeared, out of which was formed the vegetable kingdom— the celestial bodies being then set in the firmament, the sun in the day-time to keep alive and cherish the vegetation by its warmth, and the moon and stars to shed their influences by night. The waters now produced the first animated beings in the fish and fowl which they brought forth abundantly,

condition, proceeded from the Almighty's fiat, when he said :—" Let there be light and there was light," darkness previously resting upon the face of the deep.

We are thus taught by Scripture, that our earth, previous to its formation in order, was found in a shape of some kind or other, though not clearly defined. As to its *precise state* little is said ; but even from this little, we are led to the important result, that the vast and universal deep which then covered the earth, must have rested on some foundation or other; and what could that be, but on those rocks called Primary, of which granite is generally the lowest? By this, we arrive at the conclusion, that water was coeval with original granite; and will not this, in a great measure, account for the highly crystallized state in which that rock is found : as for ages it must have formed the sole bed of the ocean or " deep?"

In the Mosaic narrative, no mention is made of the creation of any of the *rocks* which formed

and these being provided with food in their respective elements ; the earth next brought forth " the cattle and creeping thing, and beast of the earth after his kind." Thus, was every thing created " beautiful and in order and very good."

the foundation of the earth as it *then* was; the reason appears obvious, for we are led to the inevitable conclusion, that they existed *previously* to the reduction of the original chaos into order, see Gen. i. 2.

When our minds have been thus familiarized with the idea, as to what state the primitive earth, (or rather the foundation rocks and the water which covered them,) was found in, we should next consider the *design* the intelligent Being may have had, in creating from the beginning such a vast abyss of waters. And here we must be led to ponder on the deep import of those words which are the conclusion of the second verse of the first chapter of Genesis; " And the Spirit of God moved upon the face of the waters." The word which in our translation is rendered " moved," in the original Hebrew signifies *brooded*, as fowls do hatch or cherish their young; which operation intimates a *preliminary character of vitality*, but excludes the notion, that the Spirit of God can mean the *wind*, for as yet a vital atmosphere like the present was not in being.*

---

* Some philosophers of the ancient and modern school of science, object to the idea of attributing to the Divine Spirit the immediate acting or operating upon the fluid chaotic

This idea of the incubation of the Spirit, or of the principle of life, which was in operation from the beginning, and which also produced the first motion, prevails, more or less, in all cosmogonics. It was by this principle or by the Spirit of God, that motion was infused into the chaos of elements, which was followed by the creation of light, and heat the cause of expansion, and lastly, by the whole harmonious circle of creation.*

Where the Spirit of God is, there must be *creative energy;* and it seems plainly deducible from Scripture, that the Almighty, who acts prospectively to *final causes*, whose " ways are in the great deep," and " who sitteth above the water-flood," was, *at that very time*, wonderfully

mixture; and consider that the expression "Spirit of God," refers to the air in its grosser form of wind, as the material agent by which the earthly mass or shell of the earth was separated from the watery medium. The original Hebrew word for that of the Divine and natural Spirit is precisely the same through the Bible.

* See Buckland's Bridgwater Treatise, vol. 1. pp. 18 to 26. —Also Catcott on Creation and Deluge, and Jones's Physiological Disquisitions. Translation of the Pentateuch by the Rev. Julius Bate ; and Inquiry after Philosophy and Theology, by the learned Mr. Spearman, author of Letters on the Septuagint.

preparing the future earth for those beings which he intended should inhabit it; though, as yet, the operations of his Spirit, were confined to the "deep," over which he *brooded*, causing it to bring forth abundantly, not as on the fifth day of creation, "the moving creature that hath *life*," but the necessary provision for *sustaining* life, and adding to the comforts of those creatures which were destined, at a subsequent period, to owe their existence to his creative wisdom, goodness, and love.

Among the minerals, it is probable, also, that common salt is an original formation, as in sea-water it is every where held in solution. There is no production of God's providence more conducive to man's existence and comfort than that of salt: and various have been the attempts made to account for its *origin*, especially, of that with which the ocean is impregnated. The subject has long perplexed the minds of philosophers; but the difficulty appears, in this instance, to arise from their unwillingness to forego what they cannot satisfactorily explain by *second causes;* and revert at once to the *great first cause* of all, who, in the beginning, created every thing, "and behold it was very good." And though no mention is made in the Mosaic account of the important production of salt, we may fairly conclude, that it then, as it now does, formed a component part of the "deep," (Gen. i. 2,) which existed before the world began, or was formed in order. As the rocks on which that "deep" must have rested, are not named among the works of the then creation, because they existed *previously*, so we may conclude, that salt also was in a pre-existent state, and coeval with original water, in which it was held in solution.

As man was *created* in a state of great purity and innocence, being placed in the garden of Eden to dress it, and permitted to eat the fruit thereof, everything *necessary* to his existence, comfort, and happiness, is mentioned, as created in the *six days* in which the Lord made heaven and earth. The *non-essentials* are not named; but after the fall of man, other things became necessary to him, such as the minerals, and metals of all kinds; yet these he could procure only by the sweat of his brow, the condition by which he was then doomed to eat his bread.

The all-wise Being, indeed, foresaw that the wants of man would increase when his condition should, by his own sinful act, be changed; these wants, therefore, he, in his boundless love to him, had provided for in a wonderful manner.

He foresaw, likewise, that man, after his apostacy, would not be content with having dominion only over all the *earth*, which God had assigned to him; but that he would also seek to extend this dominion over even the *sea* by means of his ships. The waters thereof were, therefore, previously prepared, and made buoyant, that he might float his burthens upon them. As the ocean now occupies more than two-thirds of the globe, we cannot but recognize in this also the wise provision of the Creator, that its waters should thus be salted and seasoned for preservation and the salubrity of the atmosphere.

If the objection is raised against the theory of the ocean being impregnated from the "beginning" with salt, that this mineral is never found resting on the *primary rocks;* this difficulty may be obviated by explaining, that salt only becomes insoluble when it exceeds 40 grains of salt to 100 grains of water. It must arise, therefore, that until the salt exceeds this given quantity, it cannot form a deposit, but only by the *evaporation* of the water in which it is contained.

If there was no deposition of salt remaining from the primitive ocean, and it is asked, how then were the wants of man supplied, from the period of his disobedience, when he did eat of the forbidden fruit, to the destruction of the ancient world by the flood, a space of about 1650 years? To this objection, the answer presents itself—that salt for the first time became *necessary* to man *after* he was permitted to eat, not as during the 1650 years, of only " every fruit and green herb" but when his diet was extended, and "every moving thing that liveth was given to him for meat."

In his own mind, the author can obviate many existing difficulties with respect to the origin of many of those vast beds of salt which are now found, by supposing them to have been formed by the evaporation of the *diluvial* waters, when " the Lord made a wind to pass over the earth and the waters assuaged." Thus, this most necessary mineral would be more generally spread over the earth, the number and depth of the beds, varying according to local circumstances. From calculation it is now found, that if the average depth of the sea be five miles, and it contains $2\frac{1}{2}$ per cent. of salt, if the water were entirely evaporated, the saline residue would form a stratum of salt more than five hundred feet in thickness, covering three-fifths of the surface of the globe. In another place, the endeavour has been made to account for the *alterna-*

tions of the different strata, which in some places occur with those of salt.

Rock salt, generally occurs near the feet of extensive mountain ranges. This adds probability to the opinion, that these ranges were once the boundaries of the extensive lakes and pools of salt water which remained after the ocean had retired from the present continents; the rock salt, in these situations, being formed by the gradual evaporation of these lakes and pools; and this evaporation taking place *slowly*, regular crystals would be formed. In other places, it is probable, that *subterranean fire* may have been the active agent in the formation of rock salt, by quickly evaporating the water which lay in the valleys that were formed by the volcanic upheaving of some of these mountain ranges, before the country had emerged from the ocean. In such instances, the evaporation rendered rapid by heat, the salt would be deposited in a confused crystalline mass, as it is, indeed, in many places. The frequent occurrence of anhydrousgypsum with rock salt, which is also anhydrous, favours the supposition, that the salt with which it is associated, owes its formation to the action of heat, the consolidation of both having, probably, been effected simultaneously by this element.

But it is readily allowed that two difficulties seem to present themselves unfavourable to the preceding hypotheses.—The first is:—How could the beds of rock salt be covered with earthy strata, if they were formed by the gradual or speedy evaporation of salt water? The second, —That rock salt sometimes covers the summits of the loftiest mountain ranges.

It has indeed, been observed, that rock salt generally occurs at the feet of extensive mountain ranges, and that these, probably, formed the boundaries of the salt water lakes which remained, after the country had emerged from the ocean, on the third day of creation. If, in some instances, the process of evaporation gradually took place in a warmer atmosphere, before the deluge, see p. 41 and 42, the crystallized salt that remained might *afterwards* be covered by earthy strata during the deluge, or from partial sea or fresh water inundations. If in other places the evaporation was effected speedily by volcanic fire, during the universal deluge, a fresh and successive rush of waters would cover the mineral deposit with diluvial strata, swept along by the current, till it met with a boundary in the mountain ranges. In this manner the salt would remain deeply covered by the diluvial strata, and only be brought to light by the persevering industry of man.

With regard to the mountain ranges on whose summits salt is said to exist; it seems probable that they were elevated by volcanic agency after the deluge, by which the earth's surface was then entirely changed.

The author having previously suggested that the Creator gave our present *ocean* its saltness from the "beginning," when "darkness was on the face of the deep," will not be content without attempting to prove this; and here, he does not mean to say, that the waters of the ocean do not owe some of their saltness to the salt which the rivers, rains, and other waters, dissolve in their passage through divers parts of the earth, and which are carried into it. But these causes *alone* are insufficient to account for the average quantity of salt mingled in sea water, being about $\frac{1}{30}$th part of the whole. All the rivers in the world could not have conveyed into the ocean a quantity of salt equal to $\frac{1}{500}$th part of its weight; though we suppose they daily carried into the ocean, and there left, the same quantity of salt which they at present daily deposit in it. If there are, as some suppose, beds of rock salt situated at the *bottom* of the ocean, and its waters, ever since the creation have been exerting their dissolving power on such solid masses, why are they not at this time perfectly saturated with salt?

In researches made by the Honourable Mr. Boyle, to ascertain the saltness of the sea at its surface, and at greater depths, he says:— " that equal bulks of water taken up in the English channel at the surface, and at the depth of fifteen fathoms, were equally heavy, as they contained the same quantity of salt."

In other places it has been observed, that water is more salt near the *surface*, than at greater depths. This was also the opinion of some of the ancient philosophers. From experiments made on this subject, on the occasion of the voyage to the North Pole; it was found that equal quantities of water, taken up in the open sea off Shetland, 60° N. lat., at the surface, and at the depth of 65 fathoms, yielded by evaporation equal quantities of salt: namely, $\frac{1}{29}$th part of the whole.

In N. lat. 65° water taken from the surface gave nearly $\frac{1}{28}$th of its weight of salt:—at the depth of 64 fathoms, it yielded by evaporation equal quantities of salt, namely, nearly $\frac{1}{29}$th part of its weight: an equal weight of water taken up from a depth of 683 fathoms in the same place, gave only $\frac{1}{32}$nd of its weight in salt. From these results, we find that water is more salt near its *surface*, than at greater depths, which would not be the case were the bottom of the sea stored with *inexhaustible* rocks of salt, which some imagine.

Where, in some instances, the saltness of the sea has been found to be greater at a lower depth, it may arise from the discharge of *fresh water*, which being lighter than the sea water, will not readily mix with it; and therefore, the superficial water being most readily diluted, will contain less salt, than that which is at a greater depth. It may also owe this difference, in *some few instances*, to the beds of fossil salt, which it is possible, *may* be situated at the bottom of the ocean; but the difference, generally, between the saltness of the surface of the sea, and the water at a greater depth, may be more satisfactorily accounted for, by the exhalation of vapour: for, in all the experiments that have been made, it has been found that where the heat of the sun is greatest, the surface of the water is more salt; the purer parts of the element being extracted and drawn up into the atmosphere by the sun's influence.

From the results of the foregoing researches, it appears evident, that the ocean cannot owe its saltness to the solution of beds of mineral matter at the bottom; nor to the agency of rivers or sub-currents discharging their contents into it. These appear insufficient, satisfactorily, to account for a *general and pretty equally diffused* saline impregnation of about $\frac{1}{30}$th part of the whole; and we should take into consideration, that the fact, of the ocean *generally* containing at its bottom beds of rock salt, has never yet been substantiated. In the difficulty then which arises, in accounting for the natural production of common salt, how can we help recurring to *Him*, who is the first cause of all we know, and of what we *cannot* understand, who by his almighty fiat called light out of darkness, at the beginning of the formation of all things?

CHAPTER II.

THAT there are many natural *basins* which contain successive layers of salt, over which alternations of strata and marine remains occur, cannot be denied; but this is no conclusive evidence, against what has been advanced; namely, that the beds of salt which are found in every country, were formed by slow or speedy evaporation of the waters in which that mineral was contained. The *sea* and *land* alternations, peculiar to the basins, may easily be accounted for by *successive eruptions* of the sea, which, in latter times, filled those natural basins: but when the water was dried up by evaporation, the salt would, of course, remain at the bottom, over which, in process of time, a layer of soil might be formed, either by the decay of vegetation, or by the drying up of a *fresh water* inundation. And here we should notice, that the organic contents of basins, can furnish no correct evidence to geologists in illustration or confirmation, of any *general* theory. Every one is aware, that the nature of a basin, implies, that it is separated from the sea coast by *ridges*, so that its deposits, both of strata, and organic remains, become *peculiar*. In every separate basin they may be different, according to the various changes which have acted upon them.

Cuvier supposes, " that there has been an alternate flux and efflux of salt and fresh water over the country around Paris." May not the same have taken place in our own country?* for, at Headon Hill, in the Isle of Wight, at the isles of Purbeck, Sheppy, &c., successive formations of fresh and salt water have been discovered; these singular fossil formations seem to contain mixtures of fishes and land productions, and of salt and fresh water shells and fishes.

Would not our modern geologists more effectually establish their theories, if they prosecuted

---

* An inundation of the sea recently happened in Newry. It is thus described in the *Times* paper, Dec. 1838. "About 8 o'clock on Wednesday night, the tide rose suddenly and impetuously, and poured through the streets with awful rapidity. The middle of Hill Street was in some places two and in others three feet and upwards under water. The area in front of the Telegraph establishment was submerged to the depth of four feet; and in the counting-houses and lower offices, the water was upwards of five feet deep; so quickly did the tide tumble in, that books and papers of every kind were under water."

their researches *generally*, and in various parts of the world, instead of limiting them (as it is to be feared, they too commonly have done) to the vast natural basins in the *vicinity* of *London* and *Paris?* There are to be found intermixtures of *sea, fresh water*, and *land fossils*, in these basins, which certainly shew that they have alternately been occupied by sea and fresh water.

If we seek to discover, whether there have been successive formations of vegetable and animal life, and which of them is the most ancient, it will be necessary, that we should extend our researches deep into the crust of the earth, till we come to those rocks which are called primary or most ancient. These rocks generally occur, in immense masses, and form the lowest part of the earth's surface we are acquainted with; they constitute the foundation on which rocks of the other classes are laid.

The most ancient rocks are found to contain neither animal nor vegetable remains; for they are extremely hard, and the minerals they are composed of, are more or less perfectly crystallized: this leads us to conclude that they were formed prior to the creation of organic beings.

Having failed to discover fossil, animal, or vegetable remains in the primary rocks, the transition rocks must next be searched. These cover the primary, and here we find the first appearance of the *lowest kinds* of vegetable and organic remains. These rocks may therefore be regarded, as presenting the most ancient records of our globe's earliest inhabitants. As we ascend from the transition rocks to the more recent strata, we discover a progressive development, and a succession of more perfect forms. This appears confirmed by the fossil remains of animals and vegetables peculiar to these. Every regular stratum in which they are disseminated, was once the uppermost rock, however deep it may be below the present surface, or with whatever rocks it may now be covered. Thence, we learn, that the secondary strata were formed in succession, and thus these fossil remains preserve the records of the ancient condition of our planet.

It is only by *divine revelation*, that man can learn the state in which the earth existed at the creation, as also the order of that creation, and of the world's *earliest* inhabitants. This knowledge, our first parents must have derived from an *immediate* communication from the divine Being, and which has been transmited to us, not by tradition only, but by means of the inspired writings. The most ancient of these, are the

Pentateuch and Book of Job. The Book of Job, Lord Bacon was of opinion, is pregnant with the mysteries of natural philosophy. There is no doubt, that both these books, if applied to aright, will reflect light on the researches of modern geology.*

In the 26th and 28th chapters of Job, there are many mysteries of philosophy enlarged upon. In the 25th chapter, after Bildad had endeavoured to exhibit to Job the power of God, Job rebukes him in the 3rd and 4th verses of the 26th chapter, by ironically asking him,—" If he has *plentifully* declared God's power as it *really is*, and by whose spirit he had done so?" He then proceeds himself in the 5th verse, to declare greater instances of the works of God, than Bildad had done, for among other proofs of these, he says, " Dead things are formed from under the waters, and the inhabitants thereof:" and in the 7th verse, " he hangeth the earth upon nothing." In the 14th verse, Job, after enumerating many instances of God's power, de-

* See Pikes's Philosophia Sacra, edited by the Rev. S. Kittle; and Jones's first principles of Natural Philosophy, also, his Philosophical Disquisitions.—Bishop Horne's celebrated statement of the case between Sir Isaac Newton and Mr. Hutchinson.

clares, "that these are but parts of his ways," but adds, " how little a portion is *heard* of him, (understood,) but the thunder of his power, who can understand?"

The Almighty has permitted great light to be thrown on his ways by means of the discoveries of modern science. That the earth does hang upon nothing, being sustained and moved by the etherial fluid or expansion which fills and pervades all space, the system of central gravity clearly discovers; and the author has long since come to the conclusion, that science, if it will but bring divine revelation to its aid, can discover what is implied by the words, " dead things are formed from under the waters." Commentators are not all agreed with respect to the literal rendering of the Hebrew word " *Rephaim*," which in our English version is translated " dead things;" but the translation as we now have it, is nearly the same as the preceding one of Queen Elizabeth's time. In a copy, bearing date 1595, there is this printed marginal note to the text, —" The dead things are formed from under the waters and near unto them:"—" here Job beginneth to declare the force of God's power and providence, in the mines and deep places of the earth." In the Douay version of the Vulgate, the passage is thus rendered; "Behold the giants groan under the waters, and they that dwell with them." If we take into consideration the *context*, where the Hebrew word *Rephaim* occurs, our old English rendering of the passage seems more in accordance with what Job wished to set forth, namely, God's *power* and *providence*, not in *destruction*, but in the *creation* and *preservation* of those things which his bounteous goodness had called into existence.

If there arise doubts in our minds respecting the obscure meaning of those words of Job, " dead things are formed from under the waters," let us see if the researches of modern science will reflect light upon them; and in order to arrive at a conclusion, we must consider several relations between *cause* and the effective *design* of the intelligent Creator.

We find it so arranged by God's providence, that in the early stages of the vegetable and animal kingdom, there should be a progress from simple to more complex forms. In the animal kingdom there once existed forms, without head, heart, or eyes; and as we have now reason to believe, that animals of every species derive pleasure from the action of their organs, and from existence itself, what kind of pleasure could creatures like these enjoy, enveloped in darkness and without the power of moving their habitations, we can form no idea. Yet we know that the Lord has created nothing in vain, "for the earth is full of his goodness."

The Rev. William Kirby,* in the first volume of his work, "On the Power, Wisdom, and Goodness of God, as manifested in the Creation of Animals," &c., p. 162, remarks, "that the first plants, and the first animals, are scarcely more than animated molecules, and appear analogous of each other. These could most easily be converted into *nutriment* for the tribes immediately above them." Do we not discover, in this, the designing intelligence of the Creator? for on what could the creatures which the waters brought forth *abundantly* on the *fifth* day of creation subsist had it not been that *their* wants were *previously* provided for, as the wants of "the beast of the earth, the fowl of the air, and everything which crept upon the earth," were, at a subsequent period provided for, when God made the earth for all alike, giving them "every green herb for their meat?"

* See this learned and truly Christian Author's Introduction and Appendix to his work on Entomology—a Bridgwater Treatise.

As it is the author's belief, (and he knows it to be a scriptural one,) that nothing which was endued with life, was destroyed until *after* man's transgression; that is, as we are told, when death came into the world by his sin; so in endeavouring to prove that the earliest marine inhabitants of the deep, (if inhabitants they could be called,) were constituted to become the *food* of those higher orders which were brought into being on the fifth day of creation, it will appear a startling assertion, when he says, it does not, consequently follow, that the life of any of them was destroyed.

We know that most, if not all, the marine tribes of the present time, are predaceous. The land animals which existed before the flood, *to them* every green herb was given for meat. On what then, may it be asked, could those animals have subsisted, that lived in the sea from the period of their creation on the fifth day, to the fall of man, when death first commenced his reign in the world?—probably, on nothing but those *molecules* which had previously been prepared for their food, and which Job seems to refer to as the " dead things formed from under the waters."* It produces

---

* The words of Job "Dead things are formed from under the waters" seem to imply, that, God first formed the food

more sublime ideas of the power and goodness of God, in thus conceiving him working *silently*, by second causes, in the deep, providing at the same time for the wants of every creature inhabiting it; and it seems more in consonance with the methods he *now* adopts, to bring about those great final results, which from eternity he had predestinated "according to the good pleasure of his will."

That scientific botanist, Mr. Brown, in the course of his investigations into the mysteries of natural philosophy, has discovered that not only vegetable, but mineral molecules, when placed in a *fluid medium*, would move about in various directions; but, as to the cause by which these motions were generated, he offers no conjecture. These motions, he observed, and was satisfied, arose neither from currents in the fluid, nor from its gradual evaporation, but belonged to the

for the inhabitants of the deep, and when that was prepared the "inhabitants thereof" were then called into life, the waters being commanded to bring forth "the moving creature that *hath life*," Gen. i. 20 ; this implies that previously to the fifth day of creation, they had brought forth moving creatures *not endued with life*, for the word "*life*" seems given here in contradistinction to that which had it not, and it is remarkable that the designation " moving creature that hath life" is only applied to those tribes of fish and fowl which the *waters* brought forth.

particle itself. And from the spherical molecules mixed with the other oblong particles, obtained from *cleraki pulchella*, that they were in rapid oscilatory motion in both mineral, animal, and vegetable substances. Along with the molecules, he found other corpuscles, like short fibrils, somewhat moniliform, or having transverse contractions, corresponding in number as he conjectured with the molecules comprising them; and these fibrils, when not consisting of more than four or five molecules, exhibited motion resembling that of mineral fibrils, while longer ones of the same apparent diameter were at *rest*. These motions of the molecules he describes as being vermicular, which is progressive.

But another and a greater law of the Creator, seems to be ascertained from the experiments of Mr. Brown, whilst penetrating into the depths and mysteries of Philosophy before unexplored; which is, that all bodies, whether organized or unorganized, are formed as fibrin in the animal kingdom, by *spherical molecules* made, as it were, into necklaces, and then adhering in bundles, and that these are *the substratum of all substance.*

With respect to the *rocks* which are called granite, the *mineralogist* has *ascertained* that they are of a compound substance, made up of three distinct and simple mineral bodies, quartz, felspar, and mica, each presenting certain regular combinations of external form and internal structure, with physical properties peculiar to itself. Chemical analysis has shewn that these several bodies are made up of other compound bodies, all of which had a prior existence in some more simple strata, before they entered on their present union, in the mineral constituents of what are supposed to be the most ancient rocks accessible to human observation.

The *crystallographer* has *also* shewn, that the several ingredients of granite, and all other kinds of crystalline rocks, are composed of *molecules* which are invisibly minute, and that each of these is made up of smaller and more simple molecules, every one of them combined in fixed and definite proportions, and affording at all the simple stages of their analysis, presumptive proof, that they possess determinate geometrical figures. These combinations and figures are so far from indicating the fortuitous result of *accident*, that they are disposed according to laws the most severely rigid, and in proportions mathematically exact.

The crystalline mineral bodies occur under a fixed and limited number of external forms called *secondary.* These are constructed on a series of more simple *primary* forms, which are demonstrable by cleavage and mechanical division, without chemical analysis. The integrant molecules of these primary forms of crystals are usually compound bodies, made up of an ulterior series of constituent molecules, *i. e.* molecules of the first substances obtained by chemical analysis; and these, in many cases, are compound bodies made up of the *elementary* molecules, or final indivisible atoms, of which the ultimate particles of matter are probably composed. The particles of the several substances existing in nature, may thus deserve to be regarded as the alphabet composing the great volume which records the wisdom and goodness of the Creator.

We may illustrate the combination of exact and methodical arrangement, under which the ordinary crystalline forms of minerals have been produced, by the phenomenon of a single species, namely, the well known substance of carbonate of lime; chemical analysis divides the integral molecules of carbonate of lime into two compound substances, namely, quick lime and carbonic acid, each of which is made up of an incalculable number of constituent molecules; a further analysis of these constituent molecules, shews, that they *also* are compound bodies, each made up of two elementary substances; the lime being made up of elementary molecules of the metal calcium and oxygen, and the carbonic acid of elementary molecules of carbon and oxygen; these *ultimate* molecules of calcium, carbon, and oxygen, form the final indivisible atoms into which every secondary crystal of carbonate of lime can be resolved.

The author's object in thus introducing the discoveries of scientific men, is to shew the probability that animal, mineral, and vegetable molecules had their origin in that deep which covered the foundation of the earth from " the beginning," where they could exist without atmosphere and without light; and that a portion of them constituted the *food* of those aquatic tribes which were afterwards called into being, whilst another portion of them was employed in forming the *earth's surface*, as it was destined to appear, when the waters were to be gathered together into their decreed place.

It seems probable, that the mineral molecules of which the primitive rocks are made up, were at one time all held together in solution in the deep by the *balance of affinities;* when that balance was deranged, or so acted upon by external circumstances, *precipitation* would take place, and the primitive rocks would thus, in process of time, be formed in the hard and crystalline state in which they were destined to appear, when the waters were to be gathered together into their decreed place. It seems natural, and at the same time necessary, that, in order to form the primitive and transition rocks, the balance of reciprocal affinities thus held in solution, must be deranged by a variation in *temperature;* and what strengthens this conclusion is the fact, that the Geiser springs of Iceland, whilst in a *heated* state *retain* the flint and lime-stone contained in them; but when the water is exposed to the air, and it *loses* its heat, the flint and limestone are *precipitated,* and hard incrustations of these are formed in the course of the fountain.

A suggestion is here offered, that the period in which the balance of affinities was thus deranged, by a variation of temperature acting on the *surface* of the *primitive ocean,* was, when the Spirit moved (or brooded) on the face of the deep, Gen. i. 2.

The above hypothesis may seem somewhat inconsistent with the remarks already made, pages 9 and 10, where it is said that the primary rocks must *for ages* have formed the sole bed of the deep. In explanation of this, we would say, that when the molecules, of which these primary rocks are composed, became hard and consolidated by *cohesion*, they would form a fitting bed for the deep to rest upon; but previous to that period, they may have been held in suspension in the fluid medium by the law of the Creator, *universal gravitation*. It has also been suggested, pages 12 and 13, that sea-water, and the salt contained therein, together with the rocks on which the deep rested, were original formations being coeval with each other. The two last of these may have existed in the fluid medium for *ages* before the period given in the 3rd verse of the 1st chapter of Genesis; but an objection will here be raised, that if these were held together by central gravitation, there would be no necessity of forming a bed of rocks for the deep to rest upon. We should, however, always bear in mind, that the world was, at that stage of its existence, undergoing a state of *preparation* for the reception of man, who, in God's appointed time, was to have dominion over all the earth. Thus, when the original rocks were formed either by precipitation, or shifting of their mineral molecules, caused by a variation of temperature acting on the surface of the watery medium, these molecules, would swarm together by mutual attraction, and become a consolidated body on which the water would rest. This hypothesis involves no change of the great Creator's law, that of universal gravitation, as it remains the same.

The recent discoveries of science, having shewn us, that the primitive rocks, and the earth's surface, are made up of *molecules;* and that molecules, indeed, form *the substratum of all substance.* It is singular and very confirmatory of divine truth, that the first vegetation which had its origin on the third day of creation " before it grew" was brought forth by the *earth*, of which molecular matter it was instantly and wonderfully created? From the *same material,* " the living creature, cattle, and beast of the earth, on the sixth day, started into life, and, last of all, man himself was formed from the dust of the ground. The birds and fish, (which had their origin on the fifth day, previous to the creation of animals and man), we are told, were children of *the ocean.* In confirmation of this, we find them *similar* in their organic structure, and the elements in which they move appear analogous; they are similar also in their manner of moving, and propagating by eggs. Dr. Thomas Willis observes, that their brains and eyes are likewise very similar; their bodies, like those of the vegetable and animal kingdom, which were formed from the *earth*, are found also to be of *molecular* origin, or compound bodies made up of others, which had an existence in a more simple state. These *vegetable* and *animal* forms which are endued with life peculiar to themselves, after a while again resume their *prior* condition; " for dust they are and unto dust return," and like their parent earth, they also become the nourishment of worms. Thus there is a tendency manifested throughout the vegetable and animal kingdom to flourish and decay, one body becoming in succession merely the nourishment of another living one. In this uncertain state, " there is an end of all perfection." May we seek then, more *simply* and diligently after what is truth, for that we are assured abideth for *ever*.

CHAPTER III.

In the preceding chapter we have seen, that no inconsiderable portion of the earth's surface has been formed by *organic secretion* of the most simple forms, their fossilated remains being found many hundred feet below its present surface. And there was a time, when such immense multitudes of those animals flourished, which have the smallest number of organs and senses, that calcareous mountains of vast extent are found chiefly to consist of their remains.

The transition rocks are almost entirely composed from the exuviæ of such kind of forms. These rocks had their origin under a *tranquil* sea, as their sedementary structure indicates. The cause remains to be seen, why zoophytes of the most simple organization, occupy exclusively the lowest bed of the transition series which rest on the most ancient or primary rocks; for we do not find the remains of different classes of zoophytes, testaceous animals, of reptiles, and vegetable mamiferous quadrupeds, confusedly intermingled together, excepting in beds of *clay* and *gravel* near the *surface*, or in fragments of various rocks which have been broken down and subsequently united.

Bones of *vertebrated* animals, or such as had a *brain* or *spinal marrow*, have never been found in the lower strata, except a few species of fishes; nor have the bones of large mammiferous quadrupeds ever been discovered below the chalk formation. All, or nearly all, the instances that have been cited of animals of the higher class being found in the ancient strata, have proved on further examination to be fallacious; yet, when we consider what disturbing causes have acted on the crust of the globe, it need not appear surprising, if recent species of animals should sometimes be found buried in the lower rocks. May not these have fallen there through the *vertical fissures* caused by *earthquakes*, which at different times have broken down and dislocated the earth's solid strata?

In the lower strata, the animal remains are so delicate and regularly deposited, that we can have little doubt, that their exuviæ were embedded where we now find them;—but, in the upper secondary strata, the remains are dispersed and broken, and the animals appear to have perished by some convulsion of nature, such as that must have been, when the ancient world was destroyed by the *flood*. From these two facts, we acquire a perfect knowledge, that the beds which now form the crust of our planet, were deposited in different epochs of time, and under *different conditions* of the globe.

The author has stated his belief, that animal, mineral, and vegetable molecules, such as that highly intelligent botanist Mr. Brown describes, are coeval with *original water*, by which the *primitive and transition rocks*, and much of the *earth's surface*, were formed, the time being as described, first chapter of Genesis, when " darkness was on the face of the deep ;" creative wisdom and goodness, from the beginning, having laid the foundation of the vegetable and animal kingdom, with such as could most easily be converted into *nutriment*, by those tribes which in his own due time, he had destined should be called into life.

Mr. Brown has described a portion of the molecules when placed in a fluid medium, as having a *progressive* motion, whilst another portion varying in character from these, were at *rest ;* and though he does not offer any conjecture as to the cause of these varied motions, we know that every thing in nature, is the result of wise and determinate laws, and even the motions of a molecule are governed by the same wisdom which regulates each part of the external universe. In the waters of the sea, an infinity of animal, and vegetable molecules exist, some of them are locomotive; others fixed: these molecules *now* furnish a principal portion of the food of innumerable animals of a higher order than themselves. The fact, of a great portion of the molecules having a progressive motion when in a fluid medium, seems to indicate, it was thus intended, that they might every where be *disseminated* through the ocean; thus furnishing *food* for the numerous and various tribes of marine animals; whilst those which were by their structure formed to be at rest, in the very earliest stage of the earth's creation, would swarm together under the waters; and thus, in process of many ages, the vastly thick strata which we find chiefly composed of the *simplest* organic forms of vegetable and animal remains, would be formed by *molecular attraction*, and become hard and crystallized by long submergence. * But it could only be in the water of a *deep* and *tranquil* ocean, that such a hard and solid formation of

---

* Salt appearing to be an original substance, what has previously been stated with respect to the formation of rock salt, does not forbid the supposition that *mountainous masses* of salt were also formed by *molecular attraction*, the waters being *overcharged* with this mineral.

vegetable and animal together with mineral molecules could take place.

Having submitted the foregoing opinion as to the *origin* of the primary and transition rocks, and also much of the earth's surface, to the candid consideration of the geologist, we proceed to say, it is only from *Scripture*, that knowledge can be derived, as to the *period* and *condition* of our globe, when part of those numerous tribes of animals perished, whose remains are found in the *upper secondary strata.*

About 1650 years after the creation of the world, we are told, that it was destroyed by the flood, " for the earth was filled with violence, and the wickedness of man was very great;" but Noah and all his house were saved, " for he was righteous in the eyes of the Lord." In this mighty deluge " all flesh died that moved upon the earth, both fowl, cattle, and beast, and every creeping thing that creepeth on the earth, and every man." The remains of all these we might suppose would be dispersed, broken, and drifted into every land under the heavens, where, by the force of the waves, they would be embedded deep in the earth, or sometimes deposited on the *ancient formations* previously described. These covered with other matter, would form their own *peculiar* strata, precisely as they are found at this day. In latter times, the sea has, in many places, gained the ascendancy, and partially flooded some countries, leaving at its subsidence, likewise, its own peculiar deposits. These, again, would of course be different from those spread by the *universal* deluge; for some of those tribes would be wanting, which are now found in the strata formed at that great epoch, such as the genus of trilobites, and saurian tribe. Of necessity, these became *extinct* after the grand catastrophe, whereby the earth's surface, and climate were re-modelled and changed.

We learn from the apostle St. Peter, that the primeval globe, and its heavens, (or atmosphere,) *perished* at the deluge; by which expression less cannot be intended, than that the atmosphere, and the earth, were new mixed, so as to render the former less friendly to life and health; whence would gradually follow the shortening of human, and probably, animal life; and subject to raging storms and hurricanes, to the fury and fearful effects of thunder and lightning, and to the overflowing violence of torrents of rain, while the latter, from the breaking up, inversion, mixing, depression, or elevation, of its original strata, and the addition of new ones from animal and vegetable deposits, was rendered in many places utterly barren, and in others, much diminished in fertility. Such a change having taken place in the earth, and vast countries being essentially altered, both in the temperature and atmosphere, and productions of the soil, the *extinction* of many of the original animal forms incapable of bearing the past diluvial changes would necessarily follow.

Of the actual condition of the antediluvian world before this great revolution took place, Scripture does not afford us sufficient materials from which to draw very correct geological conclusions. Of this we are informed, that the life of man extended to a period of ten-fold greater duration than it does at present, which indicates a much greater salubrity of atmosphere; and, it is remarkable, that the organic remains of that first period of the human history, correspond with this indication.

The state of the air and of the seasons, which was so healthful for man, may readily be supposed to have been equally favourable to the nourishment of other organized existences; and if we are to look for proofs from Geology to confirm the assertion of the sacred volume on this point, we must seek for it in a greater *luxuriance* in the growth of plants and animals. Man himself, who seems not to have arrived at the period of puberty before sixty or seventy years of age, was probably of a superior stature than at present;—a conjecture which is confirmed by the existence of giants, as we are expressly assured before the flood, and for some time after it. Many of the organized existences of that period, were of much greater dimensions than are now to be found, either in the vegetable or animal kingdoms. Tropical plants seem to have spread over our temperate regions in great luxuriance of vegetation, and among animals, there are found in these regions some of immense proportions, whose species are now extinct. From these facts we have the evidence, that, the antediluvian climate was peculiarly genial, and, therefore, we need not be surprised to find that it was far more favourable to animal life, than the mingled and polluted atmosphere in which we at present exist.

It appears that, during such a genial epoch, the whole surface of Europe was densely peopled by various orders of mammalia, the numbers of the *herbivera* being maintained in due proportion by the controlling influence of the *carnivera*; —the individuals of every species, being constructed in a manner fitting each to the enjoyment of the pleasures of existence, and placed in due and useful relations to the animal and vegetable kingdom, by which they were surrounded. Thus, we may suppose, that, in the earliest ages of our earth's history, sharks, and the saurian tribe, would, by their voracity, keep in check the too great increase of the smaller marine and fresh water inhabitants, the time not being arrived when many species of them were to furnish food for man,—whilst the genus mammalia, spreading over the whole earth, would, by feeding on the plants and herbage, *keep down* their excessive redundancy, consequent on the universal genial temperature of the atmosphere; and here it is worthy of notice, that the bones of sharks, and of the gigantic saurian tribe, and mammalia, are discovered buried in the earth in every latitude; and also, (as before mentioned,) plants, which are found to be analogous to those which now grow in tropical climes.

There are many reasons for concluding that (as on the third day of creation), earthquakes were employed, as the second causes in separating and gathering the waters together unto one place, so at the time of the deluge, volcanic agency may have been exerted, in order to "*break up* all* the foundations of the great deep." In this case, many of those marine animals would perish, which were, at that time, within the sphere of their action; for it is known, that fishes and saurians, which are cold blooded animals, will perish immediately that the temperature of their water becomes changed from cold to warm. In this manner, their remains would be preserved *entire*, and, accordingly, we find that the greater number of fossil fishes present no appearance of having perished from mechanical violence, but seem to have been destroyed by some change effected in the water in which they moved. The circumstances under which those were found at Monte Bolca, appear to indicate that they perished suddenly, on arriving at a part of the then existing sea, which was rendered fatal to them by volcanic agency. The *basaltic rocks*, which are in the vicinity, lead to this conclusion; the skeletons of these fishes, lie parallel to the laminæ of the strata of the calcareous slate; they are always entire, and so closely packed upon one another, that

* The word *all* as it occurs here, implies volcanic subsidences, elevations, disruptions of strata, and even of continents, together with the subterraneous infusion of trap basalt, granite, porphery, and every kind of metallic vein.

many individuals are often contained in a single block. The thousands of specimens which are dispersed over the cabinets of Europe, are taken from one quarry. All these fishes must have died suddenly on this fatal spot, and have been speedily buried in the calcareous sediment then in the course of deposition.

The fishes of Torre d'Orlando, in the bay of Naples, seem also to have perished in a like manner. An entire shoal has been destroyed at once, at a place where the waters were either contaminated with some noxious impregnation, or overcharged with heat. In like manner, the proximity of this place to the Vesuvian chain of volcanic eruptions, offers the same cause to account for the destruction of these fishes, at a period *preceding* those intense volcanic actions, which have, at a more recent period, prevailed in this district. There is evidence of vast numbers of fishes and saurians having thus met with sudden death, and immediate burial; for their bodies are found entombed in the lias, in a state of entire preservation, by hundreds. Sometimes it happens, that scarcely a single bone or scale has been removed from the place it occupied during life. This condition could not possibly have been retained had the uncovered bodies of these animals been left, even for a few hours, exposed to putrefaction, and to the attacks of fishes and other smaller animals at the bottom of the sea.

The motive for describing the above interesting discoveries, is, to shew the *period* in which those now extinct animals were destroyed, which had their origin under a different condition of our planet, the temperature of which, being changed after the deluge, they ceased to exist, leaving their remains in the regular and unbroken order described; or from a variation of circumstance caused by the great deluge, confusedly heaped together in the diluvial deposit.

Some geologists are of opinion, that the absence of human bones in the stratified rocks, or in the undisturbed beds of gravel, or clay, indicates that man, the most perfect of terrestrial beings, was not created till ages *after* those great revolutions which buried many different orders and entire genera of animals, deep under the present surface of the earth. It is unsafe in the present dearth of evidence, to endeavour to establish such an anti-scriptural fact; for, have the vast diluvial beds of gravel and clay, and the upper strata in *Asia*, been searched by scientific men? All writers agree that this continent was the cradle of the human race. Supposing, that very extensive search *had* been made, and no human bones were discovered, even this can furnish no decisive proof in support of the opinion that men could not then have been settled in the countries, where the remains of those animals are found, which perished by one grand convulsion of nature.

The animals which were called into being on the sixth day of creation, as they increased by the propagation of their species, their pasture becoming diminished, would consequently spread themselves to seek it elsewhere: thus, in process of time, every part of the earth would be inhabited by some of them.

But man is a social being, an inhabitant of towns and cities. It is probable, before the flood came, mankind were spread over but comparatively a very small portion of the earth, and, perhaps, this was limited to the continent of Asia; so that it is not surprising, if the remains of man are not found in the diluvial deposit of other countries.

Besides, there appears some probability for the conjecture, that as the remains of man must have been less scattered than those of the quadrupeds, they would furnish food for those predaceous beasts, (especially the rodentia tribe,) which on their liberation from the Ark could more readily feed and batten upon bones, as the lion, the hyena, the wolf, and dog, &c. It seemed desirable, that some animals should thus become, as it were, the scavengers of the earth, in order that it might not long appear as a vast charnel-house to Noah and his family, after God had blessed them and said: "be fruitful and multiply and replenish it."

It appears worthy of observation, that the clean beasts which were taken into the ark by *sevens*, being in greater number than the unclean, which were taken in by *twos*, would sooner multiply, and spread, and fill the earth, subsisting on vegetation; whilst the smaller number of unclean animals would have no need to scatter in search of food as it might be met with close at hand in the carcasses with which the earth must everywhere have been strewn, these being consumed, they would then spread in search of it elsewhere, and when carrion food at length became scarce and difficult to find, they would be driven to prey on the herbiverous animals, which, by that time, would be found in every country on the earth.

In endeavouring to account for the absence of human bones in the diluvial deposits, the author would wish to avoid, as much as possible, anything like speculation on so difficult a subject; he therefore puts forth the above in the shape of a suggestion only, which another may take up and reason more profitably upon than he has done.

CHAPTER IV.

In the 104th Psalm, (which is supposed to have been written by Moses,) in the fifth verse, " the foundations of the earth" are mentioned. This passage appears evidently to refer to that period before the creation of light, given in the Mosaic record; for, at the sixth verse, the foundations of the earth are spoken of as being covered with the deep " as with a garment, the waters standing above the mountains." In the 9th verse of the 28th chapter of Job, reference is also made to the same period,—" when thick darkness was a swaddling band for it" (the deep);—these passages severally refer to the *same* period, given in the opening of the first book of Genesis, " when darkness was on the face of the deep."

Thus, by divine revelation, we are taught, that some of these mountains which would be essential to the enjoyment of both man and beast,*

---

* It is from the mountains we are told, that God " sendeth the springs into the valleys," and these " give drink to every beast of the field."

existed *before the formation in order* of this our world; which appears quite plain, if we consider the 6th verse of the 104th Psalm, taking it in connexion with those previously named. But should this not be quite clear and satisfactory, the author would again appeal to Scripture as his authority, for he can in this matter assert nothing of himself. In the sublime opening of the 90th Psalm, ver. 2, we are told, " that the Lord is God from everlasting to everlasting," and that he was thus " before the mountains were brought forth, or ever he had formed the earth or the world." Again, in Proverbs viii. ver. 25; —" before the mountains were *settled*, before the hills was I brought forth; ver. 26 : " while as yet he had not made the earth, nor the fields," &c. It appears that the writers of these passages, by divine inspiration, go back to the *very remotest* period, (*before* the formation of the earth and the world, as we *now* find them), when the *original* mountains were brought forth.

In the Mosaic account of the creation, no mention is made *how* the mountains were brought forth; but if we take Scripture for our guide, the reason for this omission will appear to us quite obvious; which is, that many of these mountains existed *previously*, but others, which were *afterwards* elevated by *volcanic agency*, date their origin from the third day of creation, when the waters under heaven were gathered together unto one place; whilst the waters of the sea were thus being fixed in their channel, to which " they hasted away at the thunder and rebuke of the Lord," and " a place and a bound was set for them, that they should not turn again to cover the earth," no doubt all this took place during a crisis of stupendous and terrible convulsion of nature, which seems plainly implied from the 7th and 8th verses of the 104th Psalm, where it is said "at thy rebuke, they fled (the waters*) at the voice of thy thunder they hasted away. They go up by the mountains, they go down by the valleys unto the place which thou hast founded for them."

* As the waters in these two passages of this Psalm, are described as *"hasting away and going up* by the mountains and down by the valleys unto the place God had founded for them," it evidently refers to the period given in Genesis i. verse 9, the third day of the creation, when the waters under heaven were gathered together unto one place. No one can with propriety apply the above description of the action of the waters to anything that took place at the *deluge*, for when those waters assuaged it was by a *gradual* process of evaporation caused by the wind which God made to pass over the earth.

Thus earthquakes may have been, *then*, (as they sometimes now are,) the second causes employed in forming many of the inequalities of the earth's surface; but the vast thickness of the stratified deposits, which in some countries compose entire mountain chains, cannot be said entirely to have originated on the 3rd day of creation, nor can we reasonably suppose they were formed within the period of about 1650 years, which elapsed from the remodelment to the destruction of the old world, the origin of these, as we have endeavoured to show, being much more remote.

Some modern geologists vainly endeavour to account for the immense depth of the beds containing *organic* remains, by the hypothesis of prolonged ages or days of creation, when numerous tribes of the lower orders of aquatic animals lived and flourished, leaving their exuviæ embedded in the strata in which they are now found.

In the 20th chapter of Exodus, we are told, that "in six days the Lord made heaven and earth, the sea and all that in them is, and rested on the seventh day." Throughout the detail of the six days of creation, (1st chapter of Genesis), we learn that the evening, and the morning, were the limitation of each day's work. This is the Jewish mode of computing the length of the day, which was reckoned from the beginning of one evening to the beginning of another evening. The light was called "day" and the darkness was called "night;" the days of creation consisting of an evening and morning, that is, an alternation of darkness and light, day being according to Moses, as *now*, the general term for the light portion, and the comprehensive term for the whole diuturnity, comprising all its light and all its darkness. According to this, we shall see, it is clearly specified, that each day consisted of 24 *hours;* precisely then the same number of hours the Almighty has been pleased to reveal to man, that he rested from his work, man was commanded to do no manner of work, but to keep the day on which the Lord rested, holy to his Maker, setting that time apart to the remembrance and worship of the Being who had called him into existence, giving him life and all things richly to enjoy. Are not then the notions of some geologists quite at variance with Scripture truth, when they assert, that it may have been *millions of ages* after the world was created before man was called into existence? Did not the Almighty, when he had perfectly finished the world for man's reception, form him the *last* of all things, in order that he might then give unto him the dominion over " the fish of the sea, and the fowl of the air, and over all the earth, and over all the creeping things of the earth?" Thus, all was made subservient to *mind*, but is it not reversing the order of God's providence and love, to suppose, that he created everything beautiful and " very good," in order that they might be possessed by dumb creatures without sense and incapable of appreciating and enjoying their beauty, or rendering homage to their bountiful Creator?*

It appears that the upheaving of some of the ancient primary rocks, and the formation also of the metallic veins, have a close connexion with the period of the great deluge, when our earth's surface underwent so vast and important a change. The same volcanic agency by which so mighty a revolution was probably effected, would operate in *expanding* the strata where

---

* A geologist of some celebrity has taken the sceptre of this earth, which by right divine belongs only to man, and placed it in the hands of *gigantic reptiles*, for to use his own expression, " they at one time attained an appalling magnitude, and rioting in the wide expanse of water, they swayed the sceptre of uncontroverted power over all other created beings.!!"

they were thickest, at the same time, *lifting them up* in solid masses from the *horizontal* position in which they must have been originally deposited, to their *present* level. Rocks thus lifted up with violence, would appear of a rugged nature, as, indeed, we find them; and the expanding and upheaving power which acted on the primary and transition rocks, would also act in disseminating through them those metalliferous veins which have originated in enormous cracks and crevices penetrating obliquely downwards to an unknown depth, and resembling the rents and chasms which are produced by modern earthquakes. It was of the utmost importance, that these veins should be disposed in such a manner as to render them accessible to man's industry, and this, by their present arrangement, is admirably attained. Can we fail here to recognize that wisdom and goodness of the Creator, which is displayed throughout all his works, who, at the time he was punishing man's transgression, yet in judgment remembered mercy, causing his beneficence to appear even in the convulsions which destroyed a wicked world?

Modern geologists have discovered, that a great majority of the strata have originally been formed under water, and from materials evidently in such a state, as to subject their arrangement to the operation of the laws of gravitation; had no disturbing forces interposed, they must have formed layers almost regularly *horizontal*, and, therefore, investing in concentric coats the nucleus of the earth. If their position had been left strictly horizontal, the inferior strata must have been buried for ever beneath the highest, and, in this case, we should have wanted that variety of useful minerals almost indispensable to the existence of man in a state of civilized society.

It is by the present arrangement of the earth's surface, Dr. Hutton endeavours to establish two facts; the one, "that there has been a vast convulsion of nature, by which rocks that have evidently been formed in a *horizontal* position, are now found standing *upright* or nearly so;" the other fact is, "that such appearances of rocks rugged and broken, cannot be explained by crystallization or subsidence, and, therefore, must have been occasioned by expansion of heat." This expansion must have been exerted at different times, being more violent in its operations when it raised the primary rocks, than when the secondary rocks were elevated.

That there are obviously two orders of mineral masses which form the surface of our globe, the aqueous and the volcanic, no one who is a practical geologist can deny. The sedementary formations contain the remains of organic beings, and occupy a large part of the surface of our continents, but here and there the volcanic rocks are found breaking through, alternating with, or covering the sedementary deposits. There is also another class of rocks which cannot be assimilated precisely to either of the preceding, and which is often seen *underlying* the sedementary, or breaking up to the surface in central parts of mountain chains, constituting some of the highest lands, and at the same time, passing down and forming the inferior parts of the earth's crust. This class, usually termed primary, is divisible into two groups, the stratified, and the unstratified. The stratified rocks exhibit well defined marks of successive accumulation, and consist of the rocks called gneiss, mica, schist, argillaceous slate, hornblende, primary limestone, and others. The unstratified or Plutonic rocks which evidently appear of an igneous origin, are composed in a great measure of granite, and rocks closely allied to granite. Both these groups agree in having a highly crystalline texture, and in not containing organic remains. The stratified rock, gneiss, is composed of the same ingredients as granite itself, and it occurs in beds. In hand specimens, it is often quite undistinguishable from granite, yet the lines of stratification are still evident; these lines imply deposition from water, whilst the passage of this rock into granite would lead us to infer an igneous origin. How can these apparently conflicting views be reconciled? It appears by no other way, than ascribing *one common origin* to each of these classes of rocks, that of the *aqueous;* but at the same time, not overlooking the *modification* which both *fire* and *water* may have had on these in producing a *like joint effect*, and especially, as exhibited in the *crystallization* of some of these rocks.

Anciently volcanic action, far exceeding any thing of the kind at present known, was probably the *chief agent* of God's providence in effecting many mighty changes; but the force of such volcanic action, may have been then, as indeed we now find it, confined to certain spaces;—and it was by beneficent convulsions such as these, that the present internal arrangement of the earth's surface was effected. This now enables man to supply his wants from its numerous varieties of mineral products, illustrating, at the same time the judgment and yet the paternal care of an all-directing providence. " Lord how manifold are thy works, in wisdom hast thou made them all, the earth is full of thy riches."

As to the manner in which the metallic veins were filled, whether from *above*, by *water*, or from *below*, by *fire*, much may be said on both sides of the question. The Huttonians allege, that the cracks have been made by *expansion*; and the Wernerians, that they have been caused by *subsidence*. The Huttonians in support of their assertion, maintain, that " the deeper you trace a vein, the wider it becomes;" but this the Wernerians deny, and say: " that veins terminate below, being in the form of a wedge, the edge of which is lowest."

As there appears a total discrepancy between these two theories of Hutton and Werner, the author's object will be, first to endeavour to prove that neither of them is correct, when taken in a *general* point of view, but that each may be right, when taken in a *limited* sense; and then, in order to get nearer the truth, and at the same time obviate many difficulties and inconsistencies in the two opposing theories, each maintaining that theirs *only* is the right one, he will endeavour to effect an accommodation between the two, by suggesting that *some* veins were filled from *above*, and *some* from *below*.

If we are to give credit to Dr. Hutton's *general* theory, that fissures were formed by the rending of rocks from the expansion of heat acting below, how shall we get over the fact of veins being found to exist in rocks nearly *horizontal?* Supposing Dr. Hutton to be correct, why are not these rocks raised by the expansion to a considerable angle, instead of being in the horizontal position in which we find them? Again, supposing, according to Hutton, that *all* veins were filled from an expansion *below*, would not the metallic veins be most numerous in the vertical or primary rocks; though metallic veins are more numerous in these rocks, it is a fact, that basaltic and limestone veins are no less so in the *newer* and *horizontal* rocks. If all veins were uniformly filled from below, they would give evidence of it by the uniform compactness of their substance; but this is not the case, for though many of them are filled with substances that exhibit the appearance of having been melted, others are found to contain not only sandstone, but clay, gravel, sand, and bowlders, materials which could never have been injected in a fused state from below.

On the other hand, if, according to Werner, mineral veins are *all* filled from *above* by metallic solutions, how will this account for coal being found reduced to a cinder several inches on each side where a granite vein has passed through it?* If the *general* theory of a sedementary deposition is to be relied on, it must be demonstrated by an exhibition of facts, which will, beyond all dispute prove the truth of it, and the objection which is brought against this theory should be set at rest, that the structure of many veins are not arranged according to the specific gravities of the substances as they ought to be, had they been filled only by deposition from water.

By proposing an accommodation between the two opposing theories of Hutton and Werner, and suggesting that the metallic veins were formed some from *above*, and some from *below*, at *different periods*, many difficulties which now perplex the minds of geologists will be removed.

Those metallic ores which occur in the form of *beds* in the primary and transition rocks, must be contemporaneous with the rocks in which they are found, the metallic and earthy minerals having been deposited at the same time, and have been

* A vein of basalt is found traversing Walker's Colliery, near Newcastle; it is of a fine hard, and unbroken nature, on an average about 13 feet thick; on each side of it, the coal is converted into coke to a considerable extent.

probably separated, by chemical affinity, during the process of consolidation, when those rocks were formed which had their origin at the bottom of a deep and tranquil ocean, at that period of the Mosaic narrative which has been frequently alluded to. Those metallic ores which are found in the form of regular strata in the secondary rocks, as well as the basaltic and limestone veins, may have been formed also by deposition at the bottom of the sea, with the rocks in which they are found, within the period of about 1650 years, between the creation and destruction of the old world. Here then are *two* periods of great extent allowed for these various formations.

It will be our endeavour to shew, not only the origin of the mineral veins, but the different periods of time in which we may be led to infer, that they were formed in the condition in which they now appear.

Miners frequently find one metallic vein crossing or cutting through another, and displacing it. In such instances, it is evident, that the vein which is cut through, must be more ancient than that which intersects it. As an illustration of this, see Fig. 1. The vein, *a a* divides and unites again, and finally branches off into small strings, *c c c;* *b b* is another vein which cuts through

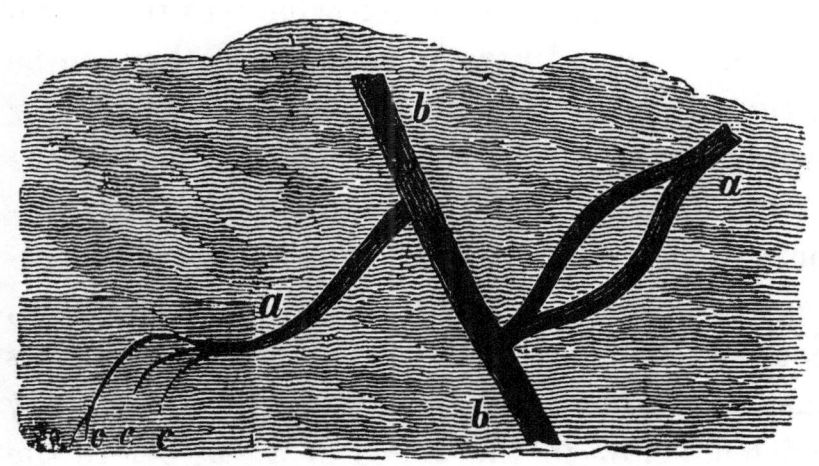

the former, and has thrown the lower part of the vein, *a a*, with the branching strings *c c c* out of its course. It is obvious, that the vein, *a a* was formed before that of *b b*, which has upheaved the rock on one side with the lower part of the vein *a*. Here then (which is very commonly the case) are two veins dissimilar in their appearance, and both evidently formed at *different* periods. It has been shewn that the vein *a a* is the most ancient, and the question now arises as to its *origin*. This, we conclude, is of an *aqueous* nature, and of course filled from *above*, as the branching off in strings downwards, seems to indicate; for had it been filled by igneous injection from *below*, those orifices would have been too narrow for the passage of such a body of fused mineral matter, which increases in width as it approaches the surface.

It now remains to be seen what was the origin of vein *b b;* and here, in reason, no other can be assigned for it, than that of an *igneous* one; for would it not require a volcanic force to displace vein *a a*, which it is utterly inadequate to attribute to water, but which a volcanic agency shooting up a mass of mineral matter from *below*, could alone effect? It seems probable, that this mass of mineral matter, must have originally existed below in the shape of a *bed*, before the formation of the vein it intersects and cuts off; this bed, probably, being formed by chemical subsidence at the bottom of that deep which the Mosaic narrative tells us, covered the ancient world from " the beginning." Previous to the fusion of this bed, and its being driven upwards in the shape of a vein, it seems probable, that the fissure *a a*, was produced by the shrinking of the materials in which the metallic vein is found, the fissure being *afterwards* filled from above by the ore in a state of *solution.** Accordingly, we find the mineral substances of which these kind of veins are composed, arranged in a regular manner in the form of *plates*, each succeeding layer being

---

* Some veins alter in character and thickness, when they leave one rock and enter another; this probably arises from a variation in the shrinking of those rocks in whose fissures the veins are now found.

deposited on the preceding. This could only be effected at the bottom of a deep and tranquil ocean, such as that might have been from the creation to the destruction of the ancient world, a period of about 1650 years. When that world, through man's sin, was destroyed by the flood, and volcanic agency most likely employed to " break up the fountains of the great deep," many of those veins which are now found intersecting others in different directions, were probably shot up in a fused state from their original beds below.

If the fact of metallic veins being frequently found to contain *different* ores at *various depths*, be brought in objection to the proposition that much of the metallic matter is a deposition from water, the author would refer again to his hypothesis of the *balance of affinities* in the waters of the primitive ocean; when, in the Almighty's providence, that balance was deranged, the materials of which the ores are composed, would, by deposition, become arranged according to their specific gravities; and thus iron ore, copper ore, cobalt ore, and silver ore, would follow each other in succession, as they actually do in some of the mines in Saxony.

In the counties of Northumberland and Durham, the deeper the lead mines are followed, the poorer they become; we can hardly reconcile this fact with the theory of injection from below, see vein, *a a*, fig. 1, which is supposed to have been formed by solution from *above*, as the branching strings indicate. The contrary is the case in the copper mines of Cornwall; these become richer, and, it is said, they are never known to come to an end, although several of them have been traced to a depth of 1000 feet from the surface; it seems reasonable to suppose, that these have been formed by injection from below, their original beds being situated at an unknown depth. But there is a general tendency of veins to take a north and south, or east and west direction, which cannot be explained by either of these theories. The interesting experiments, however, made by Mr. Fox, give some probability to the opinion, that electricity has been an active agent in the formation of these veins: but further experiments are necessary, before a theory of this kind can be fully established, and new discoveries, with respect to the nature of veins, will, no doubt, enable the miner to distinguish, in searching after the ores, which metallic veins are *entirely* of an aqueous nature, and which are of a *modified* aqueous nature, that is, modified by the action of volcanic fire. Such a knowledge, as to the origin of veins, might prove of the greatest utility.

It appears that the formation of some of the aqueous metallic veins, has a close connexion with the period, when the coal strata were deposited in the basin-shaped concavities in which they lie. In the section of a bed, Fig. 2, we see

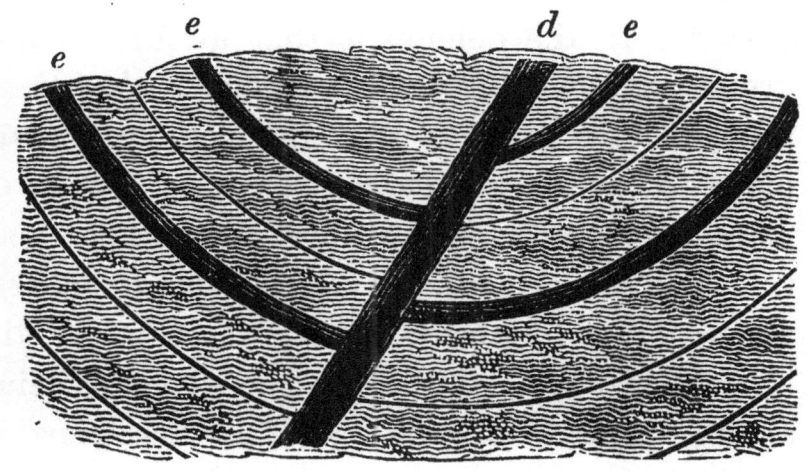

the regular coal strata *e e*, displaced and lifted up by the dyke *d*, which is composed of mineral substance, precisely in the same manner as the intersecting vein *b b*, Fig. 1, page 69, cuts off and displaces the vein *a a*. So closely are the walls or dykes associated with the beds of the coal measures, that they have been considered almost an essential part of their formation, their present condition being mainly attributable to them.

The dislocation which interrupts the progress of the metallic vein, and the regular coal strata, as shewn in the two preceding cuts, is *similar* and seems effected at the *same* period, and by the *same* convulsion. From this it appears, that the coal beds, and some of the metallic veins, were of contemporaneous origin, inclining us to give credence to the opinion of that useful and able geologist, Mr. Bakewell, namely, " that these kinds of basins in which the coal strata are deposited, were originally fresh-water lakes or marshes ; and the plants whose remains compose the coal, grew where the coal is now found, between the periodical inundations, to which the basins have been subjected. These plants being analagous to those, which at present grow in tropical climates, they must have been subjected to the same atmospheric influence which promotes the rapid growth and decay of vegetation in hot climates." Should this hypothesis of the coal formation be admitted, there seems no period more probable for the alternate formation of the coal beds, than the time which elapsed between the creation and the Noachian deluge, when the whole world and atmosphere " which then was,"

was of equable temperature.—To the same period, we may assign the formation of many of those mineral veins, which seem to be of an *aqueous* nature.

The accumulation of vegetable matter at a remote epoch in the history of the world, for the consumption of creatures which should afterwards exist on its surface, must strike the least inquiring; but when the upturned, twisted, and shattered strata so common in the districts composed of the coal measures, are before us, design is not so apparent, more particularly when the miner complains of the dislocations (faults) which interrupt his progress. We might, therefore, regard this as a bar to the ingenuity and industry of man, in extracting the combustible so valuable to him. When, however, we look more closely into this subject, we find that the shattered and contorted condition of the rocks, though it may embarrass the wishes of the miner for a time, is, in reality, highly advantageous, as preventing the passage of subterraneous waters from one mass to another, and the miners in collieries situated in one particular mass, have only to contend with the waters in it; whereas if the strata were always horizontal, unbroken, and continuous, the abundance of water that would flow into the work s would render them so difficult and expensive, that the extraction of the coal must be abandoned.

The present arrangement likewise of the materials in which the *metals* are now found, so calculated to stimulate the industry of man in search of them, exhibits no small proof of the wise foresight, and benevolent intention of the Creator, in directing, modifying, and controlling the operations of a destructive agency to *beneficial* ends. The disturbing forces by which many strata, affording numerous varieties of mineral productions, are made to emerge in succession on the surface of the earth, appear to have been effected by volcanic fires having their source within the earth itself. That there exists a mass of heated matter under the surface of the *whole* globe, seems very improbable; but that there is subterranean fire under a considerable extent of it, can scarcely be doubted. The volcanoes that are thickly scattered over both the northern and southern hemispheres, the long period of their activity, and the connexion that appears to subsist between those in distant districts,*

* During the earthquake at Lisbon, in 1775, almost all the springs and lakes in Britain and every part of Europe, were violently agitated, many of them throwing up mud and sand, and emitting a fetid odour. Even the distant

prove the depth and extent of the source of volcanic fire; but do not afford sufficient evidence in support of the theory, that there is a deep central heat under the *whole* surface of the globe.

We know that there are *very extensive regions* to which both volcanoes and earthquakes seem chiefly to be confined; these, Mr. Lyell informs us, appear to be connected with immense internal fissures in the crust of the earth, through which the central heat finds its way to the surface. He particularizes the whole region of the Andes, including Mexico and the West Indies, which pursues a course of several thousand miles, from north to south. The region of at least equal extent, which commences on the north with the Aleutian Isles, and extends, first, in an easterly direction, for nearly 200 geographical miles, and then southward, without interruption, throughout a space of between 60° and 70° of latitude to the Molluccas, where it branches off in different directions. And, in the Old World, the region extending from the Caspian Sea to the Azores, a distance of about 1000 geographical miles.

waters of the lake Ontario, in North America, were violently agitated at the time. These phenomena offer proofs of there being subterranean communications under a large portion of the globe.

"Over the whole of the vast tracts alluded to," says this author, "active volcanic vents are distributed at intervals, and most commonly arranged in a linear direction. Throughout the intermediate spaces, there is abundant evidence, that the subterranean fire is at work continuously; for the ground is convulsed, from time to time, with earthquakes, gaseous vapours, especially carbonic acid gas, are disengaged plentifully from the soil. Springs often issue at a very high temperature, and their waters are very commonly impregnated with the same mineral matters, which are discharged by volcanoes during eruptions."

The fact, then, seems to be established, that there is a vast region of excessive heat in the centre of the earth; the crust of the globe being rent internally into very extensive fissures, along which, that formidable agent approaches nearer the surface, and through which, it finds occasional or permanent vents; but there are many difficulties accompanying the theory, that this central heat is extended under the *whole* surface of our globe; one of these objections as noticed by Professor Sedgwick, is here given: "If," says he, " the earth has during any period undergone any considerable refrigeration, it must have also undergone a contraction of dimensions, and also, as a necessary consequence of a well-known mechanical law, an acceleration round its axis; but direct astronomical observation, proves, that there has been no sensible diurnal acceleration during the last 2000 years, and, therefore, during that long period, there has been no sensible diminution in the mean temperature of the earth. This difficulty does not, however, entirely upset the hypothesis of central heat; it only proves, that the earth had reached an equilibrium of mean temperature, before the commencement of good astronomical observations."

The direct astronomical observations, however, which prove that there has been no sensible diurnal acceleration of the earth round its axis during the last 2000 years, completely upsets the hypothesis of M. Cordier, who maintains, " that there is a source of immense heat in the earth, and that the external crust may be from 50 to 100 miles in thickness, and that all within this crust is a mass of melted matter: that originally the whole globe was an entire mass of melted matter, before the external crust became solid by throwing out its heat into space; and that, in this manner, the solid crust is constantly *growing thicker*, and the internal heat diminishing!"

It appears, that M. Cordier has drawn the

above conclusions, from the results of numerous experiments made by himself and others in mines and Artesian wells. With respect to the experiments made on the temperature both of the air, the water, and the rocks in *mines*, the result has been that in many cases a considerable increase of heat has been indicated by an increase of depth, though the amount remains to be proved.

In those instances, where the temperature of mines has been found to increase with the descent, we can attribute this to no other cause but to volcanic fires situated beneath these mines, which in some remote and disturbing epoch, have exerted their agency in disseminating through the rocks those metalliferous veins which are now accessible to the industry of man : but this may be one of those *partial operations* of subterranean fire peculiar to the Neptunian system which admits several of the phenomena accompanying it, which strike us most forcibly.

It can scarcely be expected, that an inquiry relating to the temperature of the *central parts* of our planet, can be brought within the limits of human observation and experiment, as the depth to which we can explore by boring or by excavation, bears so inconsiderable a proportion to the diameter of the earth. It has been calculated, that the depth of the sea in any part does not exceed 30,000 feet, or a little more than 5 miles; this, compared with the diameter of our earth, about 8,000 miles, may be regarded as nothing. That there is a vast hollow, contained in its centre which must be the seat of volcanic fire, appears conclusive, from the existence of volcanoes in every latitude of the globe. This volcanic fire, probably arranged *as a series of vaults* beneath the crust of the earth, has, in ancient times operated, as before stated, in the *mining districts*, and also in effecting the elevation of the vast *mountainous chains*, particularly of those at whose base thermal waters are found, which evidently indicate, that these are situated over or near one common source of heat, by the agency of which, the beds of the mountains have been thrown up in a situation nearly vertical.

It is not improbable, that there exists an analogy between the temperature of the Thermal waters, and the height of the hills or mountains in the neighbourhood of which these warm springs are found; as it would require a greater volcanic force to elevate the Alps mountains, (among which numerous Thermal springs are found), than it would those of our own country.*

* Thermal springs are of frequent occurrence on the summit of the Alps; these being further removed from the source

The following will shew an account of the temperature of the Thermal waters of England, and of a few celebrated Thermal waters on the Continent.

	Farenheit.
Bristol	74°
Matlock	66
Buxton	82
Bath	112° and 116
Vichy (Auvergne)	120
Carlsbad (Bohemia)	165
Aix-la-Chapelle (Flanders)	143
Aix-les-Bains (Savoy)	117
Leuk (in the Haut Valais)	117° to 126
Barèges (South of France)	120

From the experiments of M. Cordier in ascertaining the temperature of Artesian wells, it appears, that, in some parts of France, the average increase of heat above the mean temperature of the surface, is about 1 degree of Farenheit's thermometer for every 45 feet in depth; but this is liable to variation of increase or decrease in different situations.

It seems that we cannot arrive at any definite conclusion as to the temperature of Artesian wells,

of the deep volcanic heat, their temperature would, consequently, be less.

which will go to establish the theory of there being a deep central heat situated under the whole surface of our earth, unless the results of *general* researches in England, France, Italy, and Germany, are *compared together.* It is known that the southern and central departments of France, where the experiments of M. Cordier were conducted, have been the seat of active volcanoes at no remote geological epoch. This is evident from the basaltic rocks and the numerous hot springs which remain; it is therefore not improbable, that the increase of temperature with the increase of depth of the Artesian wells in that part of France, may be derived from volcanic heat situated in comparatively a quiescent state deep under the primitive or lowest rocks.*

* The elevation of temperature may, for aught we know, be confined to the neighbourhood of uplifted chains of mountains; it may be a consequence of those great natural events to which are owing the disturbance these experienced; and, consequently, it may not extend to the great plains of Russia, Siberia, Poland, and Prussia, where no such local influences exist. It is certain at least that throughout these vast tracts of level country, neither basaltic rocks, nor thermal or carbonated springs, have been noticed; whilst both the one and the other appear to become more and more abundant in proportion as we find other indications of volcanic action.

In order fully to establish the theory of a *general* central heat, Artesian wells should be sunk *in plains* remote from the scene of volcanic disturbance, by which the originally horizontal strata have been tilted up and dislocated. Any researches made in *valleys* would not present an equally fair criterion for drawing conclusions, as these valleys themselves may have been formed by volcanic *subsidence*, the internal heat still existing deep within the bosom of the earth, the nearer to which an approach being made by boring, the mean temperature would consequently be found to increase.

In England many borings for water have been executed; but it does not appear that experiments have been made on the water to ascertain the temperature. At Boston, in Lincolnshire, water was bored for, to the depth of 600 feet; the boring, during the whole depth, was in clay; and the experiment was unsuccessful, no good water being obtained. It is to be regretted, that the temperature of the water, at that depth, had not been ascertained.

The author would here hazard the following opinion, to which the foregoing observations have been tending. That the only means of coming to a correct decision with respect to the long contested theory, of there being *a general central heat* under the surface of our earth, can be by means of boring *Artesian wells*, not only in plains through clay, and the alluvial deposits, but, also, through the newer rocks of sandstone, gypsum, and shale, which are found in an *horizontal* position, or nearly so. These not having been subjected to volcanic disturbance would offer a fairer test for the experiment, than those localities which from their appearance are known to have been subjected to volcanic *upheaving* or *subsidence*.

It has been stated, that by the experiments which have been made, the temperature of some *mines* (though as it will afterwards appear not all) rises as we sink deeper into the earth. But there is a very great difference in the rate of this increase in different mines. It may be worth while to collect this rate as deduced from the different mines observed.

Huel Abraham	1° for	43 feet
United Mines	1 ,,	64
Dolcoath	1 ,,	77
Tincroft	1 ,,	$66\frac{2}{3}$
Cook's Kitchen	1 ,,	50
Whitehaven	1 ,,	43

Workington	1° for	42 feet
Teen	1 ,,	37
Percy Main	1 ,,	47
Jarrow	1 ,,	46
Killingworth	1 ,,	48
Beschertgluch	1 ,,	38·5
Hoffnung Gottes	1 ,,	57
Poullaouen	1 ,,	157
Huel Goat	1 ,,	48
Carmeux	1 ,,	50
Littry	1 ,,	35
Decises	1 ,,	30
Herzogenrath	1 ,,	76
Bagoslowsk	1 ,,	34
Charlieshope	1 ,,	22

From the above, we perceive that the difference in temperature is enormous. In the mine of Decises, there is an elevation of 1° for every 30 feet of descent; while in Poullaouen an elevation of 1° requires a descent of 157 feet, or more than five times as great. If we were to leave out Poullaouen and Charlieshope, as deviating too much from the rest, and take the mean temperature of all the others, we should obtain nearly an increase of 1° for every 50 feet of descent.

Can this augmentation and variation of temperature of mines be accounted for by any accidental causes, such as the burning of candles, the blasting of gunpowder, or the number of workmen employed in the mine? It seems probable, that much may be attributed to *natural*, and some also to *accidental* causes.

In the year 1819, the number of men employed in Dolcoath mine was 800, of whom a third part or 266 were always in the mine. The candles burnt, amounted to 200 lbs. every day. The gunpowder employed in blasting the rock in which the mine is situated, amounted to $86\frac{2}{3}$ lbs. per day.

Now, the quantity of water pumped daily out of Dolcoath mine was 535,173 gallons, or nearly $4\frac{1}{2}$ millions of pounds; and this water had the temperature of 84°, or 33° higher than the mean temperature of Redruth, where the mine is situated: thus, the heat carried off daily, in the water pumped out of the mine, would have heated 824,000 lbs. of water from the freezing to the boiling point.

From the facts ascertained respecting animal heat, we have reason to conclude, that the heat given out daily by 266 men, would be sufficient to raise the temperature of the whole water from the mine one degree.

200 lbs. of candles, according to the best experiments of the heat evolved during the burning of tallow, would be only sufficient to raise the temperature of the water so much as $\frac{1}{10}$th of a degree.

Thus, it is clear, that all the heat from these adventitious causes, does not amount to so much as $\frac{1}{15}$th of the 33° of heat which the water contains above the mean temperature of Redruth, which must be very nearly 51°.

In order to explain this curious fact, many Geologists have adopted the notion that the temperature of the central parts of the globe is much higher than that of the surface; so high indeed as to be in a state of intense ignition. According to this hypothesis, if the mean temperature at the surface be 56°, since the temperature augments 1° for every 50 feet, it is obvious, then, that the temperature of the centre of the globe must be almost 418,000°.

But though this supposition of a central heat existing under the surface of our earth, seems plausible, it must not be concealed, that there are several circumstances which greatly militate against it. Had this central heat really existed, the temperature of the earth would have varied by gradual refrigeration, during the last 2000 years, and, consequently, the length of the day become altered, but M. Arrago has demonstrated that, during this long period, no sensible diminution of that temperature has taken place.

It is sufficiently obvious, that the temperature of a place is regulated by the latitude. The mean temperature of the equator is $81°·5$; that of latitude $45°$ is $56°$; that of Stockholm in north latitude $60°$ is $41°$, and that of Lapland in north latitude $67°$ is $32°$. The following table will show the difference in temperature between the frigid and temperate zones, both in summer and winter:—

PLACES.	Latitude.	Mean Winter Temp.	Mean Summer Temp.
Madeira	$32°·37$	$63°·5$	$72°·28$
Italy	$40·50$	50	77
France.	$43·30$	$44·6$	$75·2$
France.	$47·10$	41	68
England	$53·30$	$37·8$	62.6
Scotland	57	$36·14$	$56·48$
Sweden	$60·30$	$24·8$	60.8
Gulf of Bothnia	$62·5$	$16·7$	59
Norway	71	$23·9$	$44·7$

It is obvious from this table, that, in all these places, the higher the sun is elevated above the horizon, and the longer it continues above it, the higher is the temperature; and that in winter, when the altitude is low, and the days short, the temperature is much lower. From all this, it cannot be doubted, that the temperature is regulated entirely by the sun. Now, how could this be the case, if there existed a central fire, which makes its influence be felt so much as to raise the temperature 33° at a depth of 200 fathoms under the surface? It has been said, indeed, that the heat lost by radiation is just equal to that transmitted from the centre, so that the surface cannot be heated by the central fire, but only by the sun. But, whatever effect may be ascribed to radiation, surely it must act *equably* on every part of the surface of the earth. But the poles are twelve miles nearer the centre than the equator is. Now, if every 50 feet of descent occasion an increase of 1°, twelve miles should occasion an increase of 1291°; so that allowing the heat dissipated from the surface by radiation to be equal at the poles and equator, still there ought to be an increase of more than 1200° of heat at the pole, derived from the central fire; and the temperature at the pole, instead of being 13°, ought to be enormously high. The low temperature of the pole, owing to the long absence of the sun, and probably also the absence of land, seems totally irreconcileable with the existence of a central fire in the globe, or, at least, that this central fire transmits heat to the surface in such quantity as to affect the thermometer.

The observations of Mr. Moyle, which were made during a series of years in Cornwall, seem to shew that the high temperature of the mines in that county, continues only while they are *working*. When these mines are abandoned, they get filled with water, which, of course, remains stagnant, and the temperature gradually sinks as it approaches that of the mean temperature of the place. The Oatfield engine shaft, at the depth of 182 fathoms, had a temperature of 77° while the mine was working. Mr. Moyle examined the temperature at that depth, a few months after the mine had been abandoned, and found it reduced to 66°. He tried it again many months after and found the temperature reduced to 54°. Thus, the temperature at the bottom of this mine had sunk, after it was abandoned, no less than 23°. It is obvious, that if the original high temperature had been derived from the central fire, the mere abandonment of the mine could not have reduced it.

Mr. Moyle found the temperature in the abandoned mines of Herland and Huel Alfred, that of the former 54°, and of the latter 56°, and this at all depths. The working of these two mines being resumed, the water was drawn off, and Mr. Moyle examined it during the observation, to the depth of 100 fathoms, without finding any increase of temperature.

Huel Trenoweth is another example which Mr. Moyle has brought forward. It is 100 fathoms east of Crenver and Oatfield mines, on the same *load*. This mine was discontinued working for more than twelve months, at least, as far as regards the presence of the miners; but the engine was still kept working to relieve the other two mines. The adit at which the water is discharged, is 32 fathoms from the surface. Here its temperature was 54°; and it gradually increased from this place to the mouth of the pump, where it was 56°; 15 fathoms deeper, the walls of the shaft were 54°. A gallery at this level, 40 fathoms east of the shaft, was only 53°; 5 fathoms deeper, or 52 from the surface, where there is a second cistern of water, the water was 57°; the walls in the same place were $54\frac{1}{2}°$. At the bottom, in 66 fathoms, the water that ran through a small crevice, as well as the walls of the shaft, were still 54°. The temperature of the air before going down was 68°; after returning, 64°. Here, it is obvious, that there was no increase of temperature for 34 fathoms, being precisely the same at the bottom as at the adit shaft.

Mr. Moyle exhibits the whole of his observations in a table which he has drawn up. They show, in the clearest point of view, that the elevation of temperature is chiefly confined to mines at *work*, and disappears when they are abandoned.

These facts seem to prove that the increase of temperature observed in mines as we descend, cannot be owing to the heat communicated from the *central* parts of the earth; for if it were so derived, it would not disappear when the workmen left it, or when stagnant water was allowed to accumulate in it.

If it were merely the air in the mine, or even the surface of its walls which augmented in temperature, there would be little difficulty in accounting for the phenomenon. The number of miners usually present in the mine at one time, (often amounting to 400,) together with the heat from the candles and gunpowder exploded, might easily be supposed to increase the temperature very considerably, and this temperature would naturally increase somewhat as the mine deepened, in consequence of the greater density of the air; but we have seen that these are utterly inadequate to account for the augmented temperature of the vast quantity of *water* which is daily pumped out of these mines.

It has before been stated, that the quantity of water pumped daily out of Dolcoath mine was 535,173 gallons; this water had the temperature of 84°, being 33° higher than the mean temperature of Redruth, where the mine is situated. It was likewise shewn, that if all the adventitious causes attending the working of mines, were put together, they would not have been sufficient to raise the temperature of the *water* $\frac{1}{15}$th of the 33° of heat which the water contained above the temperature of the place.

It likewise appears evident, that the temperature of the *air* in mines, as well as the surface of their walls is increased, in proportion with the number of men that are employed in them, and the quantity of candles and gunpowder expended in lighting and blasting; this heat may also be augmented by the chemical changes which are known to take place in the mineral substances in mines, from access to the water or the atmosphere.

But to whatever causes we may ascribe the

augmented temperature of the *air* and *rocks* in mines, the facts above stated are quite insufficient to account for the very great increase of the temperature of the *water* of those mines; we can ascribe this phenomenon to no other cause than that of a volcanic heat which still exists in those localities where the mines are found; this heat, as we have before stated, at some remote epoch, having acted in such a manner as to disseminate through the rocks those metalliferous veins which otherwise would never have been within the reach of man's industry. But this *local* volcanic action, as will be perceived, was *peculiar* in its nature and design, and in no way serves to establish the theory of a *general central heat*.

In pages 64 and 65 it has been shewn that some metallic veins owe their dissemination through the rocks with which they are associated, to *volcanic fire;* the dissemination of other veins being attributed to deposition from *water*. In every case, where the veins have been shot up through the superincumbent rocks by the action of fire, we may suppose the latent cause *still to exist* deep beneath the mines upon which it has acted.

In veins that have been formed by deposition from water, no volcanic heat having been employed, it will not appear surprising, if we do not discover an augmentation in the temperature of the air of some mines, as we descend: and it seems probable that fresh discoveries will establish the truth of what, at present, is only put forth as a theory, namely,—That, in all cases, where the temperature in mines is not found to increase the deeper the descent, such mines or veins have been formed by *deposition*. This will explain why Mr. Moyle found the temperature of the air in the abandoned mines of Herland and Huel Alfred *the same at all depths*, the former being 54° and the latter 56°, there being likewise no increase of temperature in the water though it was examined to the depth of 100 fathoms.

We have now seen the nature of those convulsions, by which the earth's surface was at one period totally submerged, and at another acted upon by volcanic agency causing many of the strata containing mineral productions, to emerge to its surface. But the question will naturally arise, can we, to these convulsions, with truth attribute the formation of *all the valleys* and depressions on the earth's surface?

Whilst any *general* theory is untenable with respect to the origin of *all* valleys, the following theories which have been maintained respecting them, when taken *collectively* each according to the *peculiarity of circumstance* will, we think, satisfactorily account for every 'valley' and depression to be met with on the earth's surface. Their formation has been ascribed—

1st. To excavation by the rivers that flow through them.

2nd. To the elevation or subsidence of part of the earth's surface.

3rd. To excavations, caused by the sudden retreat of the sea from our present continents.

4th. To excavations caused by the great deluge, or by a succession of inundations that have suddenly swept over the surface of different parts of the earth.

If practical Geologists would examine the *local peculiarities* of each valley, we think they might with reason attribute its formation to one or other of the above theories, or, perhaps in some instances, to the combined operation of several causes which have produced so great a diversity in the appearance of valleys.

It seems that we can only account through the operations caused by the great deluge for those phenomena, the fragments of rocks which are scattered over extensive plains at an immense distance from the Alpine districts, where rocks similar to those fragments occur. But, though in this manner we assign a cause by which these fragments of rocks were carried to their present position, still a difficulty presents itself; for we find in the midland counties of *England*, beds of gravel and fragments of rock scattered over hills that are not only far distant from the rocks which have supplied the fragments, but which are separated from them by deep valleys over which it is supposed that the fragments could not have been carried by any power of diluvial agency. To solve this difficulty, it has been imagined that these fragments and beds of gravel were deposited by diluvial agency in their present positions, before the intervening valleys were scooped out. Yet it seems highly probable that these valleys were not scooped out, but formed by *subsidence*, after the expansion and lifting up of the strata, and this, perhaps, effected during the great deluge. Such being the case, the fragments and beds of gravel would remain where originally deposited, for, had the valleys been scooped out by the deluge, or by any subsequent agency, it is obvious, that a force sufficiently powerful to scoop out valleys, would also have swept away those loose fragments from the surface.

In Switzerland, also, there are blocks of granite,

which have been torn from Mont Blanc and the adjacent granitic range, these lie scattered over the calcareous mountains, and in the valleys of Savoy, to the distance of 60 miles, or more, from the parent rocks, and some of these blocks have traversed the Jura into France a distance of 100 miles.

On the little Saleve, in the vicinity of Geneva, at the height of fourteen hundred feet above the valley, are scattered numerous blocks of granite of vast size, not at all water worn, and almost as fresh, as if recently torn from their parent mountains. They are of that kind of granite called Protogine, in which talc or chlorite is one of the component parts, and identical with the granite of Mont Blanc, while the Saleve, on which they lie, as well as the surrounding mountains, are calcareous. On the great Saleve adjoining, there is one block of this granite seven feet in length, at the height of 2500 feet above the valley. Saussure has remarked that these blocks are not broken or shattered, as they would have been, had they been hurled with violence from the Alps; neither do the limestone strata beneath them present any appearance of having been fractured or indented by their fall; on the contrary, the blocks lie upon the surface. Two of them rest upon pedestals of limestone, a few feet above the general level of the ground. Two hypotheses have recently been formed respecting these blocks of granite : the one, that they were thrown from the mountains by an expulsive force at the period of their elevation : the other, that the calcareous mountains have been subsequently raised with their load of granite blocks upon them. Whilst there are facts which appear to favour the first of these hypotheses, there are others which are entirely opposed to the second; for, had the calcareous mountains actually been raised with their load of granite blocks upon them, the force which thus raised the mountains, would, in all probability, have *dislodged* from their positions the loose blocks of granite also.

Saussure has noticed that there are blocks which in their passage from the Alps, appear to have taken the course of the present valleys; and, where they have been carried as far as the Jura chain, they rest at various heights on the sides of that range of mountains; he supposes that the whole of the valley of Geneva, and the valleys that run from the Alps, and all the lower mountains of Savoy, were covered by the sea at the period, when the great catastrophe took place, and that the rocks were torn off and transported by a sudden rush of waters. He further supposes that the specific gravity of the blocks, being diminished by the medium in which they were borne along, they might be carried to a great distance by the violence of the current, and deposited at considerable altitudes.

But there is a fact accompanying this theory of Saussure, which seems to render it not quite satisfactory; which is, that some of these blocks are lodged at a height of 1500 feet and more above the valleys, which we are inclined to think could not have been the case had the valleys been formed *previously* to the transportation of the granite blocks.

We have, in another place, shown that the inequalities of the earth's surface, owe their origin to deposition from water, as well as to the combined force of volcanic and diluvial action; the two last exerted at the period, when the earth's solid strata was broken up, and a devastating deluge of waters swept over its whole surface. It seems natural then that those valleys or depressions were formed by volcanic *subsidence, before the country emerged from the ocean*, and the same volcanic force which elevated some of the contiguous mountains, would detach from them immense blocks of granite, hurling them with violence to the situations where they are now lodged ; and, as this operation took place at a time before the country had emerged from the ocean, the fall of these projectile blocks would, in some instances, be broken by the water, whilst others, having been projected at an *earlier period*, before the waters had gained such an ascendency as to cover, we are told in Scripture, " the mountains by fifteen cubits upward" would fall upon the bare strata or earth. And will not this account for the broken and shattered appearance of some of these fragments ?

This hypothesis may serve to explain the reason why, on the great Saleve, neither the blocks of granite, nor the limestone on which they lie, present any appearance of being fractured or indented by their fall; and also, indicate the period when the rocks were elevated from which these masses were torn, as well as that of the mountains on which they are now lodged ; for it is evident that the mountains must be the most ancient, their structure being calcareous, which, as shown elsewhere, implies deposition from water, and that at the time of our world's earliest history.

In the valley of the Arve, in Switzerland, there

are to be seen blocks of granite resting on the deep mass of sand and rolled stones that form the bed of the lower part of the valley. Blocks of similar granite, may be seen in the lake of Geneva between that city and Thonon. The transportation of these granite blocks appears to have taken place *suddenly;* but the rounded stones and sand at the bottom of the valley on which the blocks of granite rest, appear to have been subjected to attrition* by the violent agitation of water, such as was, no doubt, caused by the great deluge, through the force of which this deep mass was swept along, and, at length, here collected, forming a bed upon which the *projected* masses were *subsequently* cast on the *volcanic elevation* of

---

* Rounded stones are called water-worn fragments; but, with greater propriety, they may be termed *rock-worn fragments,* as, in some places, traces are still to be seen on the rocks, along which they have been violently transported by the action of water,—no doubt, at a period when a large extent of the earth's surface was torn off, and the lower rocks uncovered; this being what is understood by a *denudation.*

The term " water-worn" seems more properly applicable to those smooth and rounded pebbles which being subject to the action of tides are always to be met with on the sea beach.

their parent rocks, and where they probably will remain until another revolution changes the surface of our globe.

In the foregoing chapters we have briefly alluded to the many convulsions which have affected the earth's surface; some of these, we can easily perceive, have been directed to a beneficial end, such as those by which, many of the strata containing mineral productions, have been made to emerge to the light.

In the earliest ages of the world, we may suppose these convulsions had not taken place, as the wants of man were then limited to those metals which were easily found, and which occur in considerable quantity in a native state, on or near the surface of the earth, such as gold, silver, and copper ore,* and it seems not improbable that those metallic veins which were filled by

* It appears probable that in the early ages of the world, the facility with which these precious metals were acquired, was the reason of their being called in succession the golden, silver, bronze or copper, and last of all the iron ages, for as civilization advanced, iron, the most useful of all the metals, would be searched for in mines and more generally brought into use.

The first mention that is made of gold, is in Gen. ii. 11, 12—" the gold which was in the land of Havilah was good," solution *from above*, were then more easily got at than at present, as they would not be covered by any of the diluvial strata. If, from our finite capacities, we cannot always see the beneficent design of the Creator in the convulsions that have since taken place in our earth, we may rest assured that what he has done was the fiat of *justice;* for we should ever remember, that, since the apostacy of man, the world has been in rebellion against the government of God, so that even the elements have been employed as instruments in his punishment, and the vindication of God's righteous sovereignty. These elements we are told " are yet kept in store reserved unto fire against the day of judgment and perdition of ungodly men."

Delvinus has now finished the task he imposed on himself of making out a scriptural system of Geology, not by setting himself in opposition to well-known facts and well accredited theories, but viewing Geology through the medium of divine revelation, and putting aside every thing which

being probably in its virgin state. Silver is first mentioned Gen. xiii. 2. " And Abram was very rich in silver." Brass (or copper) and iron are first mentioned Gen. iv. 22, " Tubal Cain, an instructor of every artificer in brass and iron."

appears opposed to it, he has endeavoured to march onwards to the truth, having thus found that made easy and plain, which before appeared inexplicable ; and should these short observations serve to reconcile conflicting views, and at the same time clear away any existing difficulties, he can take no credit to himself ; to God's word and to the researches of truly enlightened scientific men be all the praise.

Whilst the present edition was yet in the press, the Author had put into his hands, a book entituled " Scriptural Geology, or an Essay on the high antiquity ascribed to the organic remains embedded in the stratified rocks"—By the Rev. George Young, D. D. He regrets not having met with it before—the work exceeds not 100 pages, but it is a valuable one, and though the remarks are not all in accordance with his own, still, those on the absence of human bones in the strata, and some others, fully coincide with the views contained in this little Treatise.

THE END.

ERRATA.

Page 12, line 5, for *cosmogonics* read *cosmogonies*.
  12, note, line 3, for *earthly* read *earthy*.
  17, line 22, for *anhydrousgypsum* read *anhydrous gypsum*.
  39, line 15, for ? read .